VICTOR A. KANKE

50 CRITICAL ESSAYS ON THE PHILOSOPHY OF SCIENCE

EurAsian
Scientific Editions Ltd

Tallinn, 2022

ISBN 978-9949-7485-8-7

Victor A. Kanke

50 critical essays on the philosophy of science. – Tallinn: EurAsian Scientific Editions Ltd, 2022 – 169 p.

The present edition examines the main provisions of modern philosophy of science in the form of short sketches. In particular, the author pays attention to the principle of theoretical representation, the nature of scientific theories, scientific methods, intertheoretical relations, the structure of modern science, positivistic, critical-rationalistic, analytical, phenomenological, hermeneutic and poststructuralist projects of science. A critical assessment of the achievements and omissions of outstanding representatives of the philosophy of science is also given.

The philosophy of science serves as a means of enriching the conceptual and methodological content of any branch of science. The book is intended for everyone who strives for a deep understanding of the main trends in the development of modern science and modern civilization in general.

ISBN 978-9949-7485-8-7

The book is published in the author's edition without any changes.

Image: Freepik.com

www.eurasian-scientific-editions.org

© EurAsian Scientific Editions OÜ, Tallinn, Estonia
ISBN 978-9949-7485-8-7

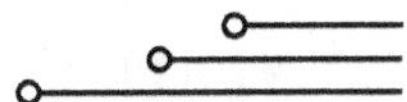

Table of contents

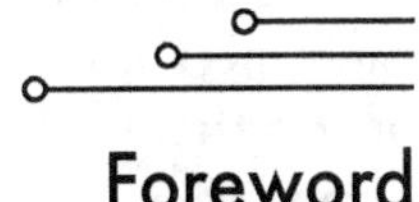# Foreword

Every time another monographic research comes to an end, I want to present its results in the most understandable form, avoiding the influence of extraneous factors of all kinds. The requirements for monographic research, in particular for a thorough substantiation of each conceptual move and its confirmation by numerous references to the works of other authors, are so strict that these studies inevitably turn into large stories.

Extensive scholarly texts are certainly relevant, but they have certain drawbacks. In my opinion, the main one is that the context of ideas often overshadows their extraordinary content. Meanwhile, it is these ideas that deserve special attention. I aim to make them stand out in their purest form, freeing up the contextual space around them.

Replacing large stories with short, concise essays is not easy. Each of them is a rethinking of some significant fragment of a large monographic study. I am sure that this rethinking has a special independent meaning, because in its absence the most relevant and interesting remains – at least partially – unknown.

This book is the result of such a "rethinking" of my encyclopedia "The Encyclopedia of Metascience and Special Philosophy of Science" (Tallinn and Wanchai: EurAsian Scientific Editions Ltd, 2021). I should also point out that my philosophical essay was largely initiated by contacts with graduate students. When they ask a question, they expect a short answer. My essay writing experience shows that each essay requires significant research effort. This circumstance testifies to the independent importance of writing an essay.

An essay is not just an arbitrary and optional addition to the main text; it usually contains at least one very unusual idea. Its development is available only to far from ordinary scientists. Even the most famous of them rarely reach the conceptual heights that with time will turn into the plots of new original essays.

All the essays in this book are critical, because in cognition, each step forward presupposes a criticism of the previous state of the theory. In the absence of criticism, essay writing is simply impossible. I can only hope that the readers of the book will appreciate – of course, with some criticism – my attempt to present modern philosophy of science in the form of essays. I am convinced that essay writing should take its rightful place in modern philosophy of science.

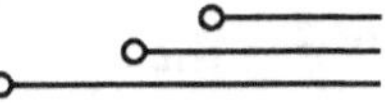

Chapter 1.
Philosophy of Science
(generalizations)

1. Philosophy of science is not a discipline, but a branch of science

What is the nature of the philosophy of science to which I devote my critical notes? The prevailing opinion is that philosophy of science like the philosophy of knowledge, society, nature, culture, and metaphysics is one of the philosophical disciplines. I suppose that this opinion is deeply mistaken. A necessary and sufficient condition for the selection of disciplines in any branch of science is that they have some common principle. For example, all physical disciplines contain the principle of least action. Significantly, the decisive events in identifying this principle's paramount importance happened in the middle of the 18th century thanks to de Maupertuis, Euler and Lagrange's research. We can expect that philosophers, like physicists, also identified a certain principle, which is the basis for distinguishing between disciplines, this time not physical, but philosophical. Unfortunately, on this score, I could not find direct indications in the works of philosophers.

Experts in the philosophy of science known to me usually associate its appearance with the rise of Carnap's positivism and Popper's critical rationalism. This means that already Plato and Aristotle developed concepts of the philosophy of science, which acquired mature forms only in the XX century. The philosophy of science, they say, has always been and remains to this day a philosophical discipline within the framework of philosophy as a branch of science. Nevertheless, the question of the principle that makes it possible to consider the very introduction of ideas about philosophical disciplines as legitimate remains unanswered.

Obviously, this principle is not included in the content of either Plato's theory of ideal forms, or Aristotle's concept of essential forms. For many centuries, philosophers have failed to find a way to the philosophical principles that make up knowledge. The situation changed dramatically after Kant proclaimed the conceptual primacy of knowledge over things at the end of the 18th century. The new trend in philosophy forced philosophers to pay special attention to theories and their principles. There is a hope that among these principles is the one, which is the basis for the selection of philosophical disciplines.

Wisest were those philosophers who linked the fate of philosophy with the successes of the sciences of nature and society, as well as formal conceptions expressing the similarity of the two classes of sciences. After many years of research, the apotheosis of which occurs only in our days, it turned out that, firstly, trustworthy knowledge is concentrated in the sciences, and secondly, that they are all guided by the same scientific methods, in particular, the methods of deduction and induction. The philosophy of science deals with the conceptual and methodological unity of all sciences. This is its distinguishing feature. There are as many philosophical disciplines as there are sciences. Philosophical disciplines are, in particular, philosophy of physics, economics and mathematics.

According to the traditional point of view, philosophy of science is a discipline within philosophy as a branch of science. According to my point of view, the branch of science is the philosophy of science. As for philosophy as such, there is no reason to continue to consider it a branch of science. Its positive content located in the field of philosophy of science. The philosophy of knowledge is the conceptual and methodological nature of scientific theories. The philosophy of society is equal to the philosophy of the social sciences, the philosophy of nature to the philosophy of the natural sciences. Metaphysics is a departure from the scientific line. It is not a philosophical discipline.

The substitution of philosophy of science as a branch of science by philosophy has an extremely detrimental effect on its fate. Many scientists sincerely would like to use the philosophical potential of knowledge. In this regard, they, as a rule, turn not to the philosophy of science, but philosophy in its traditional sense. Having failed to achieve decisive success, researchers stop using philosophy as a storehouse of current scientific achievements. They transfer the negative attitude towards philosophy to the philosophy of science as well. As a result, traditional philosophy turns out to be a barrier to the natural development of the philosophy of science. We have to admit that modern science has not yet freed itself from the unscientific content of philosophy. The separation of chaff from seeds is more painful in philosophy than in all other fields of knowledge.

2. World consists of theories

Starting my philosophical essay, I will first prove that the world consists of theories and only theories. Of course, I know that for many people, including scientists, who consider the theory to be something secondary, my thesis looks strange. Nevertheless, real philosophizing cannot do without it.

Many intellectuals, not without reason, believe that science allows judging the most thoroughly the nature of the all. Science, which consists of scientific theories, is the key to understanding everything that exists. At the same time, another thought arises; science itself has a foundation, which is all that exists. It turns out a circle in the explanation. Scientists often see the foundation of science in everything and the foundation of everything in science. Is it possible to find a way out of the described circle, if so, what is it? Many famous philosophers tried to find an adequate answer to this question. Not all of them clearly understood the nature of the contradiction under discussion. However, this does not apply to the two eminent German philosophers, Immanuel Kant and Georg Hegel. Kant proclaimed himself the author of the Copernican revolution in philosophy.

Up to now it has been assumed that all our cognition must conform to the objects; but all attempts to find out something about them a priori through concepts that would extend our cognition have, on this presupposition, come to nothing. Hence let us once try whether we do not get farther with the problems of metaphysics by assuming that the objects must conform to our cognition, which would agree better with the requested possibility of an a priori cognition of them, which is to establish something about objects before they are given to us. This would be just like the first thoughts of Copernicus, who, when he did not make good progress in the explanation of the celestial motions if he assumed that the entire celestial host revolves around the observer, tried to see if he might not have greater success if he made the observer revolve and left the stars at rest. Now in metaphysics we can try in a similar way regarding the intuition of objects. [Kant, I. (1998) [1789] Critique of Pure Reason. Translated and edited by P. Guyer and A.W. Wood. Cambridge: Cambridge University Press, 110-111.]

Kant's recipe is that one should go not from objects to knowledge, but in the opposite direction, from knowledge to objects. The weakness of this position is that the opposition of the subjective (knowledge) and objects is fixed. Meanwhile, nobody has been able to deduce from one extreme to another. Kant also did not succeed in this respect. Kant's relative failure stimulated Hegel. In an attempt to bridge the gap between the subjective and the objective, Hegel proclaimed their identity. They both represent the same world of ideas, the same totality, i.e. the absolute idea.

The programs of Kant and Hegel looked monumental. Nevertheless, they did not present their programs in a scientific form, without which the scientific community would deny them recognition. This proved beyond the power of both Kant and Hegel. Trying to present his program scientifically, Kant emphasized with great enthusiasm the importance of the principles of theories. They are the most significant concept of cognition, opening access to the world of things. Recognizing the relevance of the principles was not new to scientists. It was not difficult for them to make sure that Kant himself, in understanding the principles, does not always take into account the scientific advances already achieved. For example, he believed that time is the principle of arithmetic. In reality, the principles of arithmetic are a set of axioms, not time.

Hegel reduced the totality of ideas that express the identity of the subjective and the objective of a system of dialectical categories.

This conclusion contradicted the content of all sciences, in which dialectical categories are essentially absent. Philosophers-metaphysicians raised a topical question about the foundations of philosophizing and science, but they could not solve it. It turned out to be a tough nut to crack also for those philosophers who, in their constructions, sought primarily to take into account the achievements of science. Particularly indicative in this respect was the dispute between the representatives of neopositivism and critical rationalism, whose leaders were, respectively, Rudolf Carnap and Karl Popper.

Carnap, in a small book with the pretentious title "The Logical Structure of the World" (1928), tried to recreate the structure of science, starting with the world of sensations, experiments, and the inductive method, which allows developing the concepts of laws. This path of philosophizing did not lead to success. Nobody can derive from sensations the thoughts and actions of people. As for the induction method, it is not omnipotent. Along with it, there are other scientific methods.

Karl Popper sharply criticized attempts to free any part of science from dependence on theories. As for observation, researchers always make them in light of science.

Analytical philosophers have sought to combine the achievements of neopositivists and critical rationalists. Norwood Hanson argued that seeing is a 'theory-laden' undertaking. [Hanson, N.R. (1958) Patterns of Discovery. Cambridge: Cambridge University Press, 19.]

Willard Quine saw all objects as theoretical. *[Quine, W.V.O. (1981) Theories and Things. Cambridge, MA: Harvard University Press, 20.]*

At first glance, one gets the impression that the analytic philosophers prioritize the phenomenon of theory. This impression is misleading. The fact is that many analytic philosophers, following the pragmatist John Dewey, believe that scientific theories are instruments for solving vital problems. It is these problems that are primary, not theories.

Reading the texts of outstanding philosophers, I got the impression that they still have not decided on the initial link in philosophizing. Scientific theories are the first candidate for this place of honour. Nevertheless, they are not honoured with final approval insofar as many researchers downplay their importance. Theories, they say, are instrumental and, being a product of the creativity of people, do not do justice to the objective world.

In my opinion, there are all the prerequisites not only for the proclamation of the theoretical relativity of all statements, as Popper, Hanson, and Quine did, but also for the approval of scientific theories as to the only true starting link in productive philosophizing. Ultimately, in scientific theories, the entire human world's wealth is found, including subjective and objective, instrumental, and non-instrumental. Thus, I argue that the initial link in philosophizing is the principle of theoretical representation. Representation is multiple presentations. The principle of theoretical representation does not need any metaphysical props. As a self-sufficient principle, it is one of a kind.

3. The topicality of mature knowledge

In a previous essay, I substantiated the relevance of scientific theories, or in a word, science. What is science? Oddly enough, even brilliant scientists find it difficult to answer this question. Physicist Albert Einstein and mathematician Henri Poincare argued that science studies what is. Both meant that the experiment separates true from false. Science, they say, deals with the truth.

Alas, geniuses were wrong: they did not define 'what is'. Consequently, the concept of science remains unclear. It should also be borne in mind that science is not limited to an experiment, which is nothing more than only one phase of research.

In my opinion, many prominent philosophers did not succeed in explaining the nature of science either. Karl Popper argued that researchers could refute by experiment only scientific theories. This is incorrect because they refute unscientific theories also.

Reflecting on the nature of science, I concluded that it is wrong to try to find its absolute criterion. Another thing is essential, the ability to distinguish between top-knowledge and low-knowledge. Science is always the ultimate knowledge, top knowledge. Nonscience is the kingdom of low-knowledge. Let me give you an illustrative example. The ancient Indians explained the fall of the stone to the Earth by the invocation of the spirit of Mother Earth to the spirit of the stone. The great ancient Greek philosopher Aristotle explained the fall of stones by their desire for their natural place on Earth. A student of the modern school, following Newton, explains the fall of a stone by the action of gravitational forces. He knows how to calculate their value. A university physics teacher explaining the phenomenon of gravitation, as a rule, prefers the theory not of Isaac Newton, but Albert Einstein. The information of the ancient Indians and Aristotle about the phenomenon of gravitation is low-knowledge. We owe the scientific theory of gravitation primarily to Newton and Einstein.

However, it is impossible to establish the beginning of science unambiguously. On what basis do researchers believe that Newton, and not his prominent predecessor Galileo Galilei, is the inventor of the scientific theory of gravity? Only because Newton established the law of gravitation. Nevertheless, this criterion is not absolute. Conventionally, the beginning of science is associated with the names of those scientists who came up with some particularly effective concepts. For example, economist Adam Smith developed the concept of value, and Charles Darwin established the principle of natural selection.

The absence of an absolute criterion for determining the beginning of science is not relevant. It is important to be able to compare theories, to rank them. In this case, we certainly overcome the narrow horizons of misconceptions. Therefore, by definition, science is the ultimate knowledge. The entire history of the development of humankind shows that people will certainly strive for the ultimate knowledge. Why? Largely because developed knowledge contributes to the success of people in life in the most effective way. Dan Sperber and Deirdre Wilson have extensively covered this circumstance in the book "Relevance". They introduced the concept of "the principle of relevance".

> Human cognitive processes, we argue, are geared to achieving the greatest possible cognitive effect for the smallest possible processing effort. To achieve this, individuals must focus their attention on what seems to them to be the most relevant information

available. To communicate is to claim an individual's attention: hence to communicate is to imply that the information communicated is relevant. Communicators do not 'follow' the principle of relevance; and they could not violate it even if they wanted to. The principle of relevance applies without exception: every act of ostensive communication communicates a presumption of relevance. [Sperber, D., and Wilson, D. (1995) Relevance: Communication and Cognition. Second edition. Oxford: Blackwell, 162.]

According to the theory of relevance, positive cognitive effects in the absence of the principle of information relevance would be impossible. It is difficult to disagree with this. However, it seems to me that the principle of the topicality of mature knowledge differs from the principle of information relevance. In the centuries-old development of science, we see how less perfect knowledge finds its continuation in a more developed one. In the next essay, I will show that the transition from less mature knowledge to more mature is inevitable. This means that the principle of the topicality of mature knowledge organically expresses the nature of science. However, to show this, I will have to delve into the content of science in much more detail than has been done so far.

4. The inner life of theories

The elements of science are concepts, which in their unity form theories (conceptions). Subdisciplines, disciplines, and branches of science consist of theories. To understand the world the nature of theories is of paramount importance.

First, I note that theories contain entities. These are objects of natural theories, subjects of axiological (from the Greek αξίες [axies] – values) conceptions, and uniforms of auxiliary theories. Physics as a natural theory operates with objects, for example, particles and fields. Economics as an axiological concept deals with subjects, i.e. with individuals and groups of people guided by certain values. Auxiliary or formal theories, such as mathematics and computer science, express similarities in concepts. I suggest calling them uniforms. For example, triangles are uniforms, not objects or subjects. Entities are unity of their elementary features, properties, and relationships, in a word, variables. Examples of variables are masses and momenta in physics, prices of goods and services in economics, angles and lengths of sides in a triangle.

Along with entities, theories contain laws and principles. Variables are signs of things, their properties, and relationships. Combinations of variables form principles and laws. Examples of laws are Newton's second law in classical mechanics, the law of demand in economics, the Pythagorean Theorem in geometry. Principles, as complex concepts, differ from the laws which content they define. Examples of principles are the principle of least action in physics, the principle of maximizing expected utility in economics, and the axioms of mathematics.

Theories are not static being, but dynamic processes. In this regard, concept management techniques, namely deduction, adduction, induction, and abduction, are critical.

First, let me turn to the deduction. It is a prediction of the value of variables based on principles and laws. A deduction is always the expectation of some future.

Secondly, the researchers check the correctness of predictions by experiments, observations, and practical actions. We have to go from hypothetical variables to real ones, to facts. I called this method adduction. Adduction is attachment. In our case, we join hypothetical variables to facts. Thus, the method of facts obtaining is adduction.

Third, scientists derive the mean of variables, the laws, and principles from fact statistics. It is about induction, a method of data processing.

Fourth, in line with the results of induction, researchers are updating previously used predictive principles. The abduction method defines the rules for the realization of this operation.

Let us take a breath – the cycle of knowledge has closed (!): deduction – adduction – induction – abduction. All four methods together I call transduction (from Lat. trans – through).

Thus, the nature of scientific theory has been unravelled – it is a cycle of concepts. Everything that happens is the movement of concepts, nothing else.

Children and adults, women and men, first-graders and professors – they all "twist" concepts, but different and with varying degrees of success. The kid promises his mother to become obedient. This is a deduction. From the law "the more obedient I am, the more my mother loves me" the conclusion follows: I must become more obedient. The person makes a compliment to his girlfriend: "You are gorgeous in this dress!" Induction is evident (inference based on observation data). He who learns from his mistakes makes abduction.

Let me turn your attention to the dynamic nature of scientific theory. Everything dynamic consists of process steps, theory too. The process steps of a four-stroke internal combustion engine are the inlet of the combustible mixture, its compression, the working stroke, and the release. Process steps of vital activity of the heart are 1) isovolumic relaxation, 2) inflow, 3) isovolumic contraction, 4) ejection. Process steps or stages of the functioning of a scientific theory are prediction (expectation), obtaining facts, processing facts, and updating the initial principles. The researchers name often these stages according to their methods of implementation deduction – adduction – induction – abduction.

Since everything that exists is a representation of scientific theories, then its vitality is also a manifestation of the dynamism of theories. There is nothing more dynamic than theory.

The appearance of the four terms with the same root is not accidental. According to the "Online Etymology Dictionary" of Douglas Harper, the terms 'deduction' and 'induction' have been widely used since the 15th century. The Latin ducere means to lead, to manage. Interestingly, the Latin prefix de means, among other things, down from. From this perspective, the informal definitions of deduction and induction, respectively, as the 'top-down' (from the principle of theory to experiment) and the 'bottom-up' (from experiment to the principle of theory) transition make some sense. However, as noted above, scientists must clarify the notion of the conceptual content of the theory and the experiment. They used the term 'abduction' since the 17th century. Only in the late 19th century, thanks to Charles Sanders Pearce, it became popular in epistemology. I introduced the term 'adduction' to describe the nature of scientific theories. In my opinion, contrary to a widespread dogma, induction is the method of exclusively processing data, but not of the stage of obtaining facts. The discussed use of single-root words indicates that the stages of intratheoretical transduction form a cycle.

Interestingly, this circumstance did not pass by the attention of the ancient Greeks, who used the terms 'αναγωγή', 'προσαγωγή', 'επαγωγή', and 'απαγωγή in meaning close to the Latin terms 'deduction ', 'adduction', 'induction', and 'abduction'. The root γωγή (gogi) means some kind of production. The ancient Greeks were close to understanding theories as to the cyclical production of knowledge. The Latin tradition proceeds from the understanding of theories as the management of concepts. Of course, the views of both the ancient Greeks and Latins were far from flawless. Nevertheless, they already knew how to highlight some relevant features of scientific theories.

Unfortunately, scientists, having studied the discussed concepts for centuries, have not come to a consensus regarding many of their aspects. Some praised deduction, others induction. Few understood that deduction and induction complement each other.

The Anglo-Saxons have always attached primary importance to experimental data. In this respect, there is a lot to learn from them. In the USSR in the XX century, Soviet society for decades promoted the erroneous idea of communism, which had no practical confirmation. The Germans failed with the idea of fascism, having done planetary troubles in 12 years. In both cases, the deduction was isolated from adduction, induction, and abduction. If a person or a nation neglects scientific thinking, then this inevitably leads to numerous troubles.

The fact that one cycle of cognition goes on top of another determines the stability of life. After the correctness of mistakes, the foundation of life becomes more solid. Otherwise, society weakens and degrades.

It is time to conclude. Friends, do you want to become better? If yes, then learn to manage concepts, string the cycles of knowledge on top of each other, and use theories correctly in your life. Without this, harmonious personal development is impossible.

5. Relationships between theories

In the previous essay, I looked at the internal dynamics of a theory, i.e. intratheoretical transduction. Of course, these changes manifest themselves outside of theories. Because of intratheoretical transduction, a new situation arises in the form of two theories, the old (T1) and the new (T2). Of course, their status and the relationship between them need to be understood. Intratheoretical transduction can lead to three different results.

The first is that the two theories are incommensurable with each other.

The second result may be a slight difference between the two theories; researchers often consider them identical to each other, but this, of course, is an idealization.

Of particular interest is the third case, when the old theory is the limiting case of the new theory. In this case, scientists, firstly, define the problematic nature of the original theory, since it does not provide a correct prediction. Second, they create a new theory to update the old theory.

What is this update? First, they must free an obsolete theory from false principles. Otherwise, it will not free itself from problematic nature. The old and new theories should have the same principles.

Example. One of the principles of classical mechanics is the principle of long-range action, according to which the speed of propagation of interactions can be infinitely high. Contrary to this principle, in Einstein's relativistic mechanics, the principle of short-range action takes place: the speed of propagation of interactions is always less than the speed of massless particles in a vacuum. In connection with the noted in the updated Newtonian mechanics, the principle of long-range action must be replaced by the principle of short-range action.

Secondly, the conditions are determined under which the updated theory is an effective approximation to the new concept. These conditions are defined as the passage to the limit.

Example. The lower the speed of movement of bodies, the closer classical mechanics to relativistic.

Thirdly, the new and updated theory together form the concept of league-theory.

Example. The T2 → T1{T2} league-theory combines the new T2 theory with the updated T1{T2} theory. The notation T1{T2} means that the T1 theory is considered (as indicated by curly brackets) in the light of the T2 theory. Examples of league-theories are the Feynman-Einstein-Maxwell electrodynamics and the Marx-Riccardo-Smith theory of labour value.

After the explanations made, it is obvious that two methods, namely, methods of interpretation and ordering regulate the life of successive theories. First, the interpretation of the content of an obsolete theory from the standpoint of a more developed theory update it. Then the new theory and the renewed old theory form a league-theory. In this case, an ordering of theories takes place. Thus, in the case of successive theories, two methods of intertheoretical transduction complement the four methods of intratheoretical transduction. In the light of the developed theory of intertheoretical relations, the mistakes of far from ordinary philosophers are obvious. Alas, even great philosophers sometimes make mistakes.

The mistake of Thomas Kuhn and Paul Feyerabend. According to the Kuhn-Feyerabend thesis, all theories are incommensurable with each other. Of course, some theories are truly incommensurable. Thus, it is impossible to combine a physical theory with a biological or economic conception. Nevertheless, along with incommensurable theories, there are also commensurable concepts. The Kuhn-Feyerabend thesis contradicts the possibility of a limiting transition from a more developed theory to a less developed one. Comparable with each other, in particular, is relativistic and classical mechanics, the theory of imperfect and perfect economic competition, molecular and classical genetics. In the scientific community, Kuhn-Feyerabend thesis has never found wide support.

Niels Bohr's mistake. Bohr believed that according to the principle of correspondence, the new theory should take into account all the achievements of the outdated theory. At first glance, this requirement is quite appropriate. However, this opinion did not lead to successful applications of the correspondence principle in scientific research. Researchers can say something intelligible about the relationship between the old and the new theory only after the implementation of scientific methods. It turns out that the new theory is superior to its predecessor, its limiting case. The methods of conceptual transduction lead to this conclusion, not the correspondence principle.

Karl Popper's mistake. In my opinion, Popper is the founder of the theory of intertheoretical transduction. He prioritized theories, insisting on criticizing them and replacing outdated theories with their more successful rivals.

When I speak of reason or rationalism, all I mean is the conviction that we can learn through criticism of our mistakes and errors, especially through criticism by others, and eventually also through self-criticism. A rationalist ... thinks that only critical discussion can give us the maturity to see an idea from more and more sides and to make a correct judgement of it. [Popper, K. (1998) All Life is Problems Solving. London and New York: Blackwell, 84.]

The credo of Popper's philosophy was to postulate the criticism of scientific theories as a method of ensuring their progress. Criticism is a method of problem possibilities and their solution. The above sounds very plausible, but only as long as there is no attempt to concretize the content of the criticism. Popper did not explain how exactly problems should be criticized and identified. An alternative to his silence is the use of conceptual transduction techniques. Addressing them does not need any supports. The postulation of methods, in particular methods of criticism, which, are not used by scientists, is nothing more than a metaphysical ruse.

Michel Foucault's mistake. In the later period of his work, Foucault never tired of emphasizing that the main method of his work is problematization as a revelation of the history of problems.

Given a certain problematization, you can only understand why this kind of answer appears as a reply to some concrete and specific aspect of the world. There is the relation of thought and reality in the process of problematization. And that is the reason why I think that it is possible to give an analysis of a specific problematization as the history of an answer – the original, specific, and singular answer of thought – to a certain situation. [Foucault, M. (2001) Fearless Speech. New York: Semiotext(e), 173.]

I would like to emphasize once again that without resorting to the methods of conceptual transduction, the content of problematization remains uncertain.

Self-criticism of the author. In trying to develop a theory of intertheoretical conceptual transduction, I have defined problematization as a scientific method for many years. I believe that during this period I was clearly in the captivity of the views of Karl Popper and Michel Foucault.

6. What happens at the junction of league-theories?

The presence of different types of league-theories raises the question of the relationship between them. How are contacts between them? To answer this question, it is necessary to single out the types of league-theories. The corresponding analysis shows that, first, it makes sense to distinguish two types of not auxiliary, but independent league-theories. We are talking about natural and axiological league-theories. In what follows, for brevity, I will sometimes use the term 'theory' instead of the term 'league-theories'. Strictly speaking, we are talking about league-theories.

Natural theories deal with natural phenomena. These are the physical, chemical, geological, and biological theories. Nature, not people determine the existence of these phenomena. This circumstance indicates their independence, including, their principles, laws, and variables.

Axiological theories deal with the human world, his values. These are for example technological, agrological, medical, economic, and sociological theories. People determine these conceptions. Nevertheless, they do not depend on the natural sciences. In this sense, they are also independent conceptions.

Along with independent theories, there are also dependent theories. This means that they have an auxiliary (but not secondary!) meaning. Such theories are formal. These are philosophical, linguistic, logical, mathematical, informational, psychological, pedagogical, ethical, legal, and political theories. A characteristic feature of these theories is that they deal with similar aspects of the independent theories. Modern philosophy of science deals with the conceptual structure of theories. Linguistics distinguish parts of speech and writing. Logic masters the phenomenon of conclusions. Mathematics deals with equations and inequalities. Computer science designed to develop machine calculations. Psychology defines the individual characteristics of people. Pedagogy form adapted theories. Ethics establishes what should be. Jurisprudence defines the boundaries of acceptable. Political science examines the phenomenon of power.

The division of scientific labour leads to the existence of dependent theories. Representatives of independent theories resort to the help of other scientists. In particular, physicists turn to mathematicians, computer scientists, and philosophers of science for help. Economists also need the help of ethicists, political scientists, and lawyers. Each independent theory is hung with bunches of dependent theories.

For what follows, it is essential to clarify the content of that similarity that determines the content of dependent theories. Consider two linear relationships, one from physics, the other from economics.

The physical equation of Newton's second law connects force (F), mass (m), and acceleration (a): $F = m \times a$. (1)

The economic equation combines the supply volume (QS) with the price (P) and the coefficient at the price (d): $QS = d \times P$. (2)

The two equations under consideration are special: one operates with physical, and the other with economic concepts.

Mathematician expresses the similarity of two equations using the equation $y = k \times x$. (3) Equation (3), unlike equations (1) and (2), has a formal character. It operates with uniforms.

The quantitative relationships typical of uniforms are equally true for special equations.

After the explanations made, the meaning of the appeal of representatives of independent theories to the achievements of theories, for example, in mathematical modelling, is quite clear. This sense consists in the fact that efforts are being made to develop tendencies of the own content of independent theories. The relationship of independent theories, for example, biological and physical conceptions, is fundamentally different.

The biologist has to reckon with the reality of physical processes. How exactly does he do it? He sees physical theory as a symbol of biological theory. What does it mean? The biologist assigns to each physical factor a certain value of biological relativity. For example, he evaluates physical radiation in terms of biological changes. The biologist is not interested in physical processes as such, but in the biological changes, they cause. A physicist is not interested in biological processes in themselves, but in the physical changes that they cause. The economist directs his attention not directly to technological processes, but to the economic changes that they cause.

The recent events initiated by the coved-19 virus quite clearly indicate the specificity of the symbolic interaction of independent theories. The pandemic has caused negative phenomena in many areas of human activity. These negative phenomena indicate the symbolic significance, for example, economic or medical, of the coved-19 virus as a biological phenomenon.

We have come to extremely relevant worldview conclusions. Our world contains three components, intratheoretical and intertheoretical transduction, and interleague-theoretical relations. The latter, in the case of using the achievements of dependent league-theories, act as an intensification of tendencies that were initially contained in independent theories. The modeling method provide this intensification.

In the case of the mutual influence of independent league-theories, a special type of symbolic relativity takes place. The acceptor (independent) league-theory defines the type of symbolic relativity of the donor league-theory. The mutual influence of two independent league-theories means that each of them has the status of both acceptor and donor concepts. Physics is the donor theory from the standpoint of biology. From the standpoint of physics, biology is the donor theory.

Thus, our world is a world of independent league-theories and symbolic donor-acceptor interactions between them. Direct and reverse symbolization are methods of interdisciplinary relations. A very common mistake is the understanding of dependent theories as independent conceptions. The circle of dependent theories is very extensive.

Thus, one should distinguish between four types of intertheoretical relations. First, theories that are close in their data often are identical. Secondly, within the framework of league-theories, less developed theories are the limiting case of a more developed concept. Third, independent theories realize the potential of the dependent theories by modelling. Fourth, in the case of a correlation of independent theories, the donor theory is a symbol of the acceptor conceptions. Five methods correspond to four types of inter-theoretical relations, namely, methods of identification, passage to the limit, ordering, modelling, and symbolization.

7. People are born as potential scientists

In previous essays, I have devoted considerable attention to the nature of scientific theories. At first glance, it seems that they are available only to scientists, but not to ordinary people. Meanwhile, the remarkable similarity in the behaviour of all people is striking. As a rule, it is not difficult to establish mundane analogues of rather complex scientific concepts, in particular, deduction, adduction, induction, and abduction.

The deduction is a prediction method. All people predict the future in one way or another. Consequently, the method of deduction cannot be dispensed with here either. Often a prediction acts as an expectation of some future events. Expectation throws a person into the future.

Further. A person will certainly do something. Then he must deal with facts. An individual cannot avoid not only prediction with its method of deduction, but also obtaining facts with his characteristic method of adduction. Adduction consists of determining the values of some characteristics. No people can avoid adduction. Ask someone to answer, for example, to this question: "What did you do last night at 9 o'clock?" The answer will certainly contain the characteristics of subjects and objects, especially if you require clarification. "You said it was dark? How dark was it? ".

Induction is about making sense of facts through variables, laws, and principles. Now it is crucial to understand the nature of the relationships between the variables. Each person performs this operation. For example, someone says, "It was dark, we saw little." In this case, a connection is established between illumination and the ability to distinguish the outlines of things. There is a certain law.

Abduction gets its expression, for example, in the statement "I thought that my husband, being accused of cheating on me, would justify himself. Nevertheless, he did not do it. Now I know that he is never justified." The last sentence is a formulation of the principle of the behaviour of a woman's husband.

Consideration of numerous situations shows that all people, from toddlers to elders, schoolchildren and outstanding scientists, tirelessly carry out conceptual transduction operations throughout their lives. They do the same thing, but, of course, with varying degrees of thoroughness. By nature, all people are the same. This circumstance testifies to the fact that the scientific beginning does not represent some exoticism, which is available only to selected people. It expresses the innermost trait of all people.

It is surprising not only the striking similarity of the nature of people but also their ability to distort it. A huge number of philosophers are busy looking for an alternative to science. Often it appears as an exaltation of metaphysics. According to Comte-Carnap thesis, metaphysics is untenable because not satisfying the tests of science. The critical rationalist Karl Popper rejected this view most vigorously.

The repeated attempts made by Rudolf Carnap to show that the demarcation between science and metaphysics coincides with that between sense and nonsense have failed. The reason is that the positivistic concept of 'meaning' or 'sense' (or of verifiability, or of inductive confirmability, etc.) is inappropriate for achieving this demarcation — simply because metaphysics need not be meaningless even though it is not science. In all its variations demarcation by meaninglessness has tended to be at the same time too narrow and too wide: as against all intentions and all claims, it has tended to exclude scientific theories as meaningless, while failing to exclude even that part of metaphysics which is known as 'rational theology'. [Popper, K.R. (1963) Conjectures and Refutations: The Growth of Scientific Knowledge. London: Routledge & K. Paul, 253.]

The difficulties connected with my criterion of demarcation (D) are important, but must not be exaggerated. It is vague, since it is a methodological rule, and since the demarcation between science and nonscience is vague. But it is more than sharp enough to make a distinction between many physical theories on the one hand, and metaphysical theories, such as psychoanalysis, or Marxism (in its present form), on the other. [Popper, K.R. (1974) The Problem of Demarcation. In Popper Selections. (1985) Ed. D. Miller. Princeton, N.J.: Princeton University Press, 127.]

Popper's decisive point is that scientific theories, as opposed to metaphysical ones, are falsifiable. At first glance, Popper's argument seems to be quite consistent. However, upon closer examination, it becomes clear that it is erroneous. The fact is that the existence of a criterion for the scientific character of a theory does not indicate in any way the existence of a criterion for a metaphysical theory. The absence of a criterion for metaphysical theory and it is absent since no one succeeds in formulating it in an acceptable form testifies to the inconsistency of the term "metaphysical theory". By itself, it is devoid of any meaning. Considering the nature of metaphysical theory, we have to do this in the light of ideas of scientific theory.

Many philosophers are providers of metaphysical theories. Quite indicative in this respect is Aristotle's hylomorphism, according to which a thing, be it an atom or a man, is a unity of matter and form. In modern sciences, there are no forms or matters therefore hylomorphism is unscientific. Persisting in the desire to find out the meaning of hylomorphism, one can consider it as a semblance of scientific theories. You can compare, for example, hylomorphism with Newton's three laws, operating with a dozen concepts, including mass and force. By equating matter with mass and form with force, we can say that Aristotle anticipated the basic content of Newtonian mechanics. From the only correct, namely the scientific point of view, metaphysical theory acquires meaning only insofar as it is similar to scientific theories.

Of course, the question arises about the legitimacy of comparing metaphysical and scientific theories. This comparison is indeed legitimate. The fact is that metaphysical concepts do not arise by themselves but at the initiative of people guided by certain life ideas. A detailed analysis reveals the connection between metaphysical concepts and scientific concepts. It testifies to the possibility of comparing metaphysical and scientific theories. Refer, for example, to the work of Wilhelm Leibniz, a remarkable mathematician, and physicist and at the same time the author of the metaphysical concept of 'monad'. The monad clearly bears the stamp of differential forms so close to Leibniz as a mathematician. It is not the fruit of only fantasies far from science. Unfortunately, a full-fledged scientific rehabilitation of metaphysical concepts is always impossible.

Metaphysics is always the product of unscientific inventions. An attempt to defend its independence leads to the exaltation of these fabrications. This is the conceptual fall. People are born as potential scientists, but they constantly forget about it.

8. Many-sided theories

In what form do theories exist? This question is discussed extremely rarely, meanwhile, the understanding of the nature of science largely depends on the answer to it. It is widely believed that theory consists of thoughts or abstractions. In reality, however, things are significantly different. Knowledge about everything that exists, about objects, subjects, language, mentality, and actions, provides us with theories. This means that all these phenomena are theoretical. They are precisely the forms of presentation of theories. In the primordial form, freed from all phenomena, theories do not exist. They have flesh and this flesh is a variety of phenomena. In particular, one should distinguish between object, subjective, mental, emotional, linguistic, behavioural, and practical presentation of the theory. The same theory can be spoken, sung, danced, and shown in pictures.

Leo Tolstoy presented the theory invented by him in the novel "Anna Karenina" in a language, namely, textual form. The same theory was presented in the respective ballet, opera, drawings, and cinema. The same theory exists in different kinds it has many sides. Possessing the initial unity, all variants of the theory complement each other. The diversity of the world begins with theories. Contrary to this state of affairs, philosophers, as a rule, absolutize one of the representations of theories.

Materialists, proclaiming the primacy of matter and the secondary nature of consciousness, absolutize the object representation of theories.

Critical realists, led by Karl Popper, tend to understand the theory as a collection of thoughts.

The positivists, in particular Ernst Mach and Rudolf Carnap, reduced theories to feelings. This means that they were adepts at the sensory representation of the theory.

The mental (emotional) presentation of theories is characteristic of modern phenomenologists following the traditions of the philosophizing of Edmund Husserl.

In the 20th century, many philosophical trends insisted on the need for a language turn. This means that the representatives of these directions have absolutized the language representation of the theory. In this regard, the initiatives of the analytic philosopher Ludwig Wittgenstein and the fundamental ontologist Martin Heidegger are especially often recalled. The preference given to the language representation of the theory is characteristic of the absolute majority of modern analytical philosophers, hermeneutists, and poststructuralists.

Many philosophers put human actions at the forefront. If their value content is denied, then they are equated with behaviour. The behavioural representation of the theory is characteristic of the supporters of behaviourism, in particular, for John Watson. This movement was very popular in psychology in the first half of the 20th century. Many philosophers, notably Ludwig Wittgenstein and Willard Quine, supported it.

The practical presentation of the theory is characteristic of the pragmatists, the heirs of the philosophy of Charles Pierce and John Dewey. Unlike behaviourists, pragmatists emphasize the value-based nature of human actions. The practical presentation of theories is also characteristic of the Marxists, who, following Karl Marx, insist on the primacy of the collective objective activity of people. They prioritize the activist view of the theory.

Thus, we have to admit that the modern philosophical community ignores the unity of various representations of theories, their diversity. This is not surprising insofar as the paramount importance of the very phenomenon of the theory is not realized. By attaching instrumental significance to a theory, it is difficult to conclude that everything that happens is a manifestation of theories. One false course of knowledge entails another. The neglect of the unity of representations of the same theory leads to another incongruity. For each side of human life, they are looking for a special theory, in particular for language, thoughts, practice, and emotions. This means that the search is in the wrong direction. Theories differ in their form of presentation, and not in the conceptual content. This circumstance makes it possible critically evaluate the distinction between tacit and explicit knowledge proposed by Michael Polanyi.

> An art which cannot be specified in detail cannot be transmitted by prescription, since no prescription for it exists. It can be passed on only by example from master to apprentice. [Polanyi, M. (1958) Personal Knowledge: Towards a Post-Critical Philosophy. London: Routledge & Kegan Paul, 53.]

Polanyi believed that all information is rooted in tacit knowledge. People are not able adequately to articulate this knowledge by verbal means, therefore, "[we] can know more than we can tell". (Polanyi, M. (1966) The Tacit Dimension. Chicago: University of Chicago Press, 4.) According to Polanyi, the various forms of presentation of theories are fundamentally different from each other in the degree of them possible articulation. Practical knowledge defies articulation to the same degree as verbal knowledge. There is always the knowledge that is not properly coded.

In my opinion, some arguments destroy Polanyi's position. First, any knowledge can be refined. Secondly, various forms of knowledge representation are translatable into each other. Thirdly, the meaning of any representation of knowledge, in particular, verbal and practical, is expressed in the same way, through the same concepts. The third argument is decisive. The unity of the theory's concepts does not allow distinguishing between their meanings. Undoubtedly, they differ from each other, but only in form, and not in meaning. Polanyi mistakenly believed that knowledge differs both in form and in its conceptual content.

9. Fallibility is not a principle of human activity

To err is human. All knowledge is fallible and therefore uncertain. ... Our aim as scientists is objective truth; more truth, more interesting truth, more intelligible truth. We cannot reasonably aim at certainty. Once we realize that human knowledge is fallible, we realize also that we can never be completely certain that we have not made a mistake. [Popper K. (1996) Search of a Better World: Lectures and Essays from Thirty Year. London and New York: Routledge, 4.]

A theory, which is not refutable by any conceivable event, is non-scientific. Irrefutability is not a virtue of a theory (as people often think) but a vice. Every genuine test of a theory is an attempt to falsify it or refute it. [Popper, K.R. (1963) Conjectures and Refutations: The Growth of Scientific Knowledge. London: Routledge & Kegan Paul, 36.]

The quotes by Karl Popper testify to his search for the fundamental principles of epistemology. It is in this connection that he attaches paramount importance to the thesis about the incomplete flawlessness of any statement and, accordingly, theory. Popper is well aware that before him many philosophers, in particular sceptics, as well as, for example, Charles Sanders Pearce, the author of the term 'fallibilism', emphasized a certain uncertainty of the theory. Nevertheless, they did not draw the proper conclusions from this statement, in particular, that any theory is worthy of criticism and, ultimately, falsification.

In my opinion, Popper's argumentations contain a serious flaw. People do make mistakes at times. Nevertheless, this circumstance does not exhaust their activity as scientists. Other characteristics of human activity may be more relevant than the presence of errors in their theories. Popper makes a fundamental conclusion without delving into the mechanism of scientific activity. In this case, only a miracle can save from metaphysical errors. As you can see, it turns out that the scientific status of Karl Popper's conception of falsificationism is not obvious, perhaps even questionable. Not all actions and judgments of people are wrong; many of them are sufficiently accurate and flawless. According to Popper, the goal of scientific activity is to increase the truth, and for this, it is necessary to eradicate errors.

Oddly enough, Popper, who certainly understood the relevance of theories, nevertheless, attached only instrumental significance to them. They are necessary to overcome problems; scientists improve them for the sake of multiplying the truth. In reality, theories are self-sufficient. Problems and the realm of truth are generated by the theory itself and do not precede it. The foundations of conceptions are in the theories themselves and they consist of the management of concepts through the methods of intratheoretical and intertheoretical transduction. Consistent use of these methods leads to overcoming problems. However, this is a secondary phenomenon, not primary.

Popper's decisive mistake is that he while presenting the foundations of epistemology does not delve into the essence of scientific theories. The cart is placed in front of the horse. As a result, true scientific progress is blocked. Popper never tired of urging scientists to problematize, criticize, and falsify theories. Unfortunately, he was not able to explain exactly how these processes should be carried out. Several of Popper's proposals seem to contradict the assertion of mine. "Propose theories, which can be criticized. Think about possible decisive falsifying experiments – crucial experiments." (Popper, K.R. (1974) The Problem of Demarcation. In Popper Selections. (1985) Ed. D. Miller. Princeton, N.J.: Princeton University Press, 126-127.)

Popper's reference to the need for decisive experiments is certainly an attempt to take into account the internal dynamics of theories. However, it is insufficient to represent these dynamics in full. It is not surprising, therefore, that it did not prevent Popper from proclaiming falsification as the main epistemological principle. Falsification is not an alternative to conceptual transduction. Fallibility is not a principle of human activity.

In my opinion, the metaphysical trap that Popper fell into in proclaiming the principle of falsification is very revealing. Some spillover effects are cited as principles. At the same time, it is not proven that they are the principles that determine the course of phenomena. By no means is every actual position a principle.

10. What is the truth?

Since time immemorial, people have argued about the nature of truth. Why are they so attracted to this question?

The genius of antiquity Aristotle gave the classical definition of truth. Truth is the correspondence of said to the existing. Anyone who calls a blonde-haired person to be blond is telling the truth. Those who call a blonde-haired person a brown-haired person are mistaken. For 23 centuries, Aristotle's definition of truth seemed flawless. Yet it contains a significant flaw.

The point is that what exists we know only due to theory. Also, it should be borne in mind that scientists will certainly improve theories. It turns out that the concept of truth should be associated with the certainty of theory. Therefore, truth is changeable; it cannot be true for eternity. Truth is relative. There is no absolute truth.

At first glance, there are many absolute truths. The prominent German logician, mathematician, and philosopher Gottlob Frege argued that equation $2 \times 2 = 4$ is irrefutable. Here, they say, is an example of absolute truth. Alas, Frege was also mistaken. Arithmetic exists in many forms, and therefore the equation in question is interpreted in different ways. One mathematician understands by $2 \times 2 = 4$ something different from his opponent.

Despite the volatility of truth, scientists have not abandoned it. Strictly speaking, they began to consider as truth only such a statement, expressed or thought, which belongs to the most developed theory. Those who use an outdated theory are delusional. The one who use the most developed theory is not mistaken. Subsequently, scientists will come up with a new theory. Any person who does not master it deludes oneself.

Aristotle's correspondent concept of truth is based on the concept of reality. If we proceed from the concept of language, then this will lead to a different understanding of the truth. The neopositivist Otto Neurath observed in 1931 that it is possible to compare some parts of the language with others, but it is unacceptable to proceed from the pre-language position, asserting the independence of reality. He did not explain how the true language position differs from the untrue one. However, he is believed to be the founder of the coherent concept of truth, which insists on the subordination of different portions of the language.

Moving from one philosophical direction to another, each time you get a different conception of truth. The position of American pragmatists, in particular of Charles Sunders Peirce, is very indicative on this score.

> [I]f we can find out the right method of thinking and can follow it out — the right method of transforming signs — then truth can be nothing more nor less than the last result to which the following out of this method would ultimately carry us. [Peirce, C.S. (1906) Collected Papers of Charles Sanders Peirce. Hartshorne C., Weiss P., Burks A. (Eds.). Vol. 5. Cambridge MA: Harvard University Press, paragraph 5.553.]

The truth is the result of research that provides maximum practical value. Representatives of natural science, who believe that nature does not adapt to the needs of people, rarely support this understanding of the truth.

The founder of fundamental ontology, Martin Heidegger, saw the meaning of human existence in the creative manifestation of his true essence.

> Assertion and its structure … are founded upon interpretation and its structure … and also upon understanding – upon Dasein's disclosedness. Truth, however, is regarded as a distinctive character of assertion as so derived. Thus the roots of the truth of assertion reach back to the disclosedness of the understanding. [Heidegger, M. (1962) Being and Time. Transl. J. Macquarrie and E. Robinson. Oxford: Blackwell, 266.]

From this point of view, not only suggestions and judgments, as analysts believe, but also people, for example, friends, can be true.

All the above interpretations of the nature of truth are insufficient insofar as they inconsistently use the achievements of scientific methodology. From the standpoint of the theory of conceptual transduction, the developed theories make it possible to interpret the content of their predecessors. Only the theories that have not been refuted are true. It is necessary to improve, and this is impossible without new theories. The centuries-old study of the nature of truth convinces us that it is impossible to do without the separation of knowledge into more and less developed theories. Without the concept of truth, it is difficult to imagine modern science.

11. What is good? The scandal surrounding ethics

At the beginning of the 20th century, the Englishman George Moore, one of the founders of analytical philosophy, made a splash with his declaration that good is indefinable. A paradoxical situation has developed: everyone talks about the good, but no one can give it a clear interpretation. All outstanding philosophers found it difficult to define the nature of good. In order being not unfounded, I will give some views.

The genius of antiquity Aristotle wrote three voluminous books on ethics. He expressed a point of view that was very popular among the ancient Greeks. Good is the moral foundations that allow you to avoid passions both for lack and for excess. Good, for example, is one who can avoid thoughtless courage and cowardice. The ancient Greek recipe is to stay in the middle! Nothing too much!

In the 18th century, the German philosopher Immanuel Kant decided to illuminate the question of the nature of good in more detail than anyone else did before him. He concluded that everyone should act in such a way as to adequately represent humanity, in particular, not to lie, not to steal, not to kill. Frankly speaking, Kant did not have enough science. At the forefront, he put a special principle, a categorical imperative: Do so that the maxim of your will is by universal legislation. Let us put it simply: do what every person should do. Kant was sure that everyone knows what he should do. His advice was not specific enough. The correct view is that every man can learn good deeds. How exactly one should learn good deeds, Kant could not explain.

In the 19th century, two English philosophers Jeremiah Bentham and John Stuart Mill proposed a fundamentally new concept of ethics, namely utilitarianism. Bentham, wanting to be as specific as possible, insisted on the need to maximize the pleasure of people and minimize their suffering. Unfortunately, Bentham, as well as Mill, a follower of his ideas, faced great difficulties in determining the nature of pleasure and pain. Their critics have sometimes argued that utilitarianism is characteristic not only of humans but also of pigs, which also experience feelings of pleasure and suffering. Mill responded to this criticism not without a fair amount of irritation.

> It is better to be a human dissatisfied than a pig satisfied; better to be Socrates dissatisfied than a fool satisfied. And if the fool, or the pig, are of a different opinion, it is because they only know their own side of the question. The other party to the comparison knows both sides. [Mill, J.S. Utilitarianism. (2001) [1863] Kitchiner: Batocher Books, 13.]

The meaning of Mill's remark is that good is always what "is better". Good and evil are countable. What and how should be counted, Mill did not explain.

Scientists usually think that good is the main concept of a special philosophical discipline, ethics which scientific status raises great doubts. Ethics is usually absent in recognized sciences, for example, in economics. In the latter, among the hundreds of books, there is hardly one in which ethical issues are considered productively.

This is a rather complicated situation. Ethics, which is often proclaimed to be the main property of humanity, is absent in a distinct scientific form. A real scandal!

At this point, at the risk of being immodesty, I dare to state that I have managed to understand the essence of the ethical issue and propose a conception that removes all the difficulties described above.

I have always considered science to be the highest achievement of humanity. Anyone genuinely interested in the success of science should strive to register ethics at the department of science. Another observation was that no sciences of man and society, avoid ethical conflicts. Economists are quite successful in regulating cash flows, but suspiciously often past the poorest segments of the population. They use the principle of maximizing the rate of return on advanced capital, but this is not enough to assert ethical ideals. My decisive thought was that ethics is an organically part of the axiological sciences, but not in the form in which philosophers develop it. Scholars, in particular economists, sociologists, lawyers, historians, educators, and psychologists, generally reject what philosophers propose them on behalf of Aristotle, Kant, Bentham, and Mill.

I propose to introduce into all axiological sciences the principle of maximizing the prosperity of all stakeholders (interested persons). Prosperity is a concept not of philosophical ethics, but the axiological sciences themselves. Concerning economics, this means that it is necessary to maximize not only the profits of entrepreneurs but also the size of the wages of workers and employees. About nuclear energy, prosperity extends not only to the workers of the nuclear power plant, but also to the population, and not only in the 30-kilometre zone. Concerning pedagogy, we are talking about maximizing the competence of not only students but also teachers and parents of students.

Another feature of the situation under consideration is that the ethical principle should be in the first place in the structure of any axiological theory. Otherwise, ethical conflicts will certainly arise.

Unexpectedly it turns out that ethics is not a philosophical science. As a separate science, it makes sense only in the following case. Researchers extract ethical content directly from axiological sciences, generalize and develop it, and then offer it to representatives of axiological sciences for further using.

The extraction-generalization-development-return operation I described takes place not only in the case of ethics but also, for example, law and mathematics. Following them, ethics falls into the category of dependent axiological sciences.

So what is good? It is an improvement in the position of individuals and social groups, all stakeholders by the principles of axiological sciences. The ethical project with its centuries-old history is not outdated. It is the decisive antidote to all negative phenomena, in particular, wars, corruption, violence, impoverishment of the population, and its inadequate medical service. There has hardly been a greater shame in the history of humankind than the alienation of science from ethics. We must get rid of this shame!

By the way, now it becomes clear why George Moore did not reveal the nature of good. He sought goodness where there is none.

12. What is responsibility?

Many researchers, dissatisfied with the constant oblivion of ethics, insist on the ethics of responsibility, in the absence of which humankind, they say, cannot avoid the most harmful cataclysms for it. Their efforts are not in vain. Scientists have never before paid as much attention to the concept of responsibility as they do today. It is more surprising that questions remain about its nature.

German sociologist Max Weber contrasted the ethics of responsibility with the ethics of conviction.

> [T]here is a profound opposition between acting by the maxim of the ethic of conviction (putting it in religious terms: 'The Christian does what is right and places the outcome in God's hands'),

and acting by the maxim of the ethic of responsibility, which means that one must answer for the (foreseeable) consequences of one›s actions. [Weber, M. (1994) Political Writings. Translated by P. Lassman. Cambridge: Cambridge University Press, 360.]

In my opinion, Weber attributed the understanding of responsibility to the field of jurisprudence rather than ethics. He clearly shows the connection between responsibility and punishment for possible misconduct. Unlike jurisprudence, ethics is concerned not with punishment and setting the boundaries of the permissible, but with what needs to be done. What kind of person can we call responsible? Someone who does not deserve punishment, perhaps a lawyer would say. An ethicist is interested in something else: what should be a responsible person. Unfortunately, many scientists replace the ethical concept of responsibility by a legal one.

Many researchers connect the nature of responsibility with the possibility of a person exercising free will. At the same time, they believe that it is, in principle, impossible to realize free will. Consequently, the concept of responsibility is untenable. Friedrich Nietzsche promoted this position with great pathos.

No one gives a man his qualities, neither God, nor society, nor his parents, nor *he himself. No one* is responsible for existing at all, for being formed so and so, for being placed under these circumstances and in this environment. His own destiny cannot be disentangled from the destiny of all else in past and future. . . We are necessary, we are part of destiny, we belong to the whole, we exist in the whole. There is nothing which could judge, measure, compare, or condemn our being, for that would be to judge, measure, or condemn the whole. *But there is nothing outside of the whole.* [Quoted from: Bakewell, C.M. (1899). The Teachings of Friedrich Nietzsche. *International Journal of Ethics 9(3): 323.]*

The German-American ethicist Hans Jonas expressed a fundamentally different position. He elevated the concept of responsibility into a principle. Jonas is firmly convinced that ethically, it is not irresponsibility that prevailed, but a responsibility. This happened, first, due to the transformation of technology, ambivalent in nature, into a planetary factor. The very survival of man was at stake. In this situation, the principle of responsibility is, to avoid the worst, the main existential choice of a person. Philosophically, Jonas gravitated towards Husserl's phenomenology and Heidegger's fundamental ontology. His position is distinctly metaphysical, not scientific. To a much lesser extent, this applies to the discursive ethics of Karl-Otto Apel and Jürgen Habermas.

These authors believe that in the carefully prepared discourse, participants agree on their values. This just means that they are responsible to and for each other. Responsibility is the result of successful communication.

It is time for me to express my point of view. In my opinion, no authors mentioned above said the most important thing about responsibility. As before, I proceed from the theory of conceptual transduction. Responsible is that individual or that organization, which in all forms of its life use the most developed scientific theories. It is easy to see that my definition of responsibility is fundamentally different from understanding it as accountability. The instance of responsibility of an individual or a social group of people is they, and not external forces pressing on them.

As for the possibility of responsibility, no one, including Friedrich Nietzsche, refuted it. Not everything depends on individuals and social groups of people. Responsibility ethics advocates consider this. They believe that people are responsible to the extent that what they do depends on their competence. Such dependence exists and, therefore, liability is possible. The achievements of the sciences, and not the metaphysical reasoning of Nietzsche about the absorption of man by the cosmic whole, testify to the possibilities of people.

What is the responsibility, an ordinary concept or a principle? I think it is a principle. Earlier I argued that scientific ethics begins with the principle of maximizing the prosperity of all people involved in a particular situation. Responsibility is the obligation to use the most advanced theories. These two statements are equivalent to each other. If people use the most developed theories, then they implement the maximization of people's prosperity. Conversely, where people maximize human prosperity they are cultivating the most developed theories. Thus, we are dealing not with two, but with one principle. It is permissible to call it both the principle of maximizing the prosperity of people and the principle of responsibility.

13. What is beauty?

Many people believe that truth, good, and beauty are the main values of a person. Truth is a feature of the most advanced scientific theories. They, and accordingly whatever they represent, are true until are not refuted. Good is a generalized characteristic of what should be by the content of the most developed axiological sciences, i.e. sciences not about nature, but people. What is beauty? In the novel by F.M. Dostoevsky's "The Idiot" Ippolit Terentyev ascribes to Prince Myshkin the statement "Beauty will save the world." Is it so? Should we agree with the striking statement of William Edward Burghardt Du Bois?

I am one who tells the truth and exposes evil and seeks with Beauty for Beauty to set the world right. [Du Bois, W.E.B. (2003) The Wisdom of W.E.B. Du Bois. Ed. By Aberjhani. New York: Citadel Press, XI.]

In previous essays, I have repeatedly emphasized that the most developed knowledge of all things is contained in the sciences. It must be assumed that the nature of beauty should be discussed based on certain sciences. Nevertheless, the fate of beauty science is amazing. The philosophical doctrine of beauty emerged only in the middle of the 18th century. It was the aesthetics of the German philosopher Alexander Baumgartner. The sciences of beauty emerged even later, in the 19th and 20th centuries. I mean numerous concepts of art, in particular, the theory of painting, graphics, architecture, literature, cinema, theatre, opera, stage, circus, choreography, and ballet. Thus, the habitat of beauty is art history and its comprehension in aesthetics.

All the theories I have enumerated are scientific because they demonstrate the pinnacle of knowledge. Unfortunately, many people did not rightly understand this circumstance. It is widely believed that art history deals not with concepts, but with feelings. This is a clear delusion. Feelings are a kind of concepts. We perceive the phenomena of art by our art history theories.

Beauty and its peak, the beautiful signify certain values and preferences of people. Beauty does not appear by itself; people create and improve it. Beauty is not a natural phenomenon. It is impossible to be born beautiful; you can be born with forms that correspond to some modern ideas about beauty. I do not think that a girl with the forms of the Greek goddess of beauty Aphrodite would have won a modern beauty contest. The ideas of female beauty have changed. They are now different from those of Botticelli, Titian, Giorgione, Veronese, Velazquez, and Bouguereau, who in different years created the image of Venus on canvas.

The most developed art criticism theories testify to what beauty is. Their connoisseurs' judge beauty based on a system of complex concepts. Saying briefly beauty is what people desire according to the most advanced art history theories.

Of course, not all people are adept at these theories. They often judge beauty based on common sense theories; beauty is the harmony of the object and subject with the world. Many scientists who consider it possible to judge beauty not by aesthetic, but for example, mathematical and physical conceptions, hold views that are more intricate. Mathematicians and physicists often interpret beauty as simplicity of theory, unity of principles, the symmetry of equations, the orderliness of conclusions, and their coherence.

Of cause the transfer of concepts from one area to another, for example, as in our case, from art history to physics, is a sign of bad scientific taste. It would be quite reasonable if mathematicians, physicists, chemists, and other representatives of non-art disciplines to talk not about beauty, but the perfection of their favourite theories.

In my reasoning about the nature of beauty, there is one rather piquant place. All art history sciences, under whose department I prescribed the concept of beauty, deal with fiction. It turns out that there is no beauty in the real world. Nevertheless, we are talking, for example, about beautiful girls, flowers, and sunsets. There seems to be a contradiction, but this impression is deceiving. The point is that art criticism theories are not isolated from other theories. If someone views the world from the standpoint of art theory, then he has aesthetic relativity for him. When we characterize real phenomena as beautiful, then, in fact, we are talking about their aesthetic relativity. Things in the real world are not beautiful in and of themselves. We recognize them as beautiful only insofar as they reflect the concepts of art history.

Let me now turn to the question of whether beauty will save the world. Ethics testifies to what should be in the world. In this capacity, it is also relevant for art history. It is no coincidence that we place high ethical standards on the heroes of novels, theatrical performances, films, and other works of art. Nevertheless, one should not forget that art ethics is the ethics of the fictional world. It becomes involved in real affairs when real ethics is a symbol of art ethics. The unity of the two ethics contributes to the correct reorganization of society. People design art sciences to promote the development of beauty in the world. Anyone who says that beauty will save the world is absolutizing the meaning of art history. Strictly speaking, art history is neither better nor worse than other sciences. Of course, this does not exclude a particularly reverent attitude of many people to the phenomenon of beauty. In my opinion, Du Bois belonged to these people.

Leo Tolstoy contrasted beauty and good.

> Good, beauty, and truth are put on one height, and all three concepts are recognized as basic and metaphysical. Between there is nothing of the kind. The concept of beauty not only does not coincide with good but rather the opposite of it, since good, for the most part, coincides with the victory over predilections, beauty is the basis of all our predilections. [Tolstoy, L.N. (1985). Sobraniye sochineniy (Collected works). Moscow.: Khudozhestvennaya literatura, Vol. 22, 78. (In Russian).]

There is no reason to agree with the famous writer. Good and beauty both ensure the perfection of man. To corruption of people leads not beauty, but its substitution by the ugly.

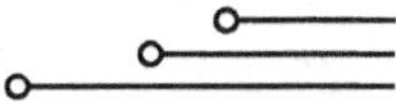

Chapter 2.
The Main Projects
of the Philosophy of Science

14. What have the positivists taught us?

There are various projects for the philosophy of science. Each of them has strengths and weaknesses. Without their criticism, it is hardly possible to understand the true content of the philosophy of science. The philosophy of science did not take on mature forms at once. The positivism with its three stages of development was the first of them. The conceptual leaders of the first, second, and third stages of positivism, respectively, were Auguste Comte and John Stuart Mill, Richard Avenarius and Ernst Mach, Rudolf Carnap and Hans Reichenbach. The third stage of positivism we know under the names of neopositivism and logical positivism.

Comte proclaimed the primacy of scientific knowledge over any other.

> In the final, positive state, the mind has given over the vain search after Absolute notions, the origin and destination of the universe, and the cause of phenomenon, and applies itself to the tudy of their laws, – that is, their invariable relations of succession and resemblance. Reasoning and observation, duly combined, are the means of this knowledge. What is now understood when we speak of an explanation of the facts is simply the establishment of a connection between single phenomena and some general facts, the number of which continually diminishes with the progress of science. [Comte, A. (1975) [1830-1842] *Cours de Philosophie Positive.* 2 vol. Paris: Hermann, vol. 1: 72.]

Mill developed the inductive method as a way of finding laws and cause-effect relationships. His original grandiose project was to give all sciences an inductive interpretation. The turn of deduction came only after induction. Mill failed to implement this project applied to political economy and geometry. He considered these sciences, in contrast, for example, to chemistry, to be deductive.

Avenarius took the experience as a starting point for cognition in the form in which it is directly cognized by people. Therefore, the scientific method consists of a pure description of the empirically given. Mach, refining this position, paid special attention to sensations as the basis of cognition. This basic is not interactions of objects and subjects, but complexes of sensations.

> For us, therefore, the world does not consist of mysterious entities, which by their interaction with another, equally mysterious entity, the ego, produce sensations, which alone are accessible. For us, colors, sounds, spaces, times, . . . are provisionally the ultimate elements, whose given connexion it is our business to investigate. [Mach, E. (1996) [1886] The Analysis of Sensations. London: Routledge and Thoemmes Press, 29.)

Carnap in the first period of his creative work, following Mach, saw the foundations of science in sensory forms of cognition. Later he abruptly changed the focus of his work. Now he saw these foundations in linguistic forms of cognition, built on the models of syntactic and semantic probabilistic logic. In this regard, he does not act as a real positivist who must give a positivist interpretation of logic. He imposes a logic on the philosophy of science.

Reichenbach, like Carnap, also paid considerable attention to logic. However, he is less an adherent of logicism than his comrade. Reichenbach research formula was "Do not pursue philosophy as an isolated science, but in the closest context with the other sciences." This approach allowed him to consider the philosophical questions of physics with great success. Reichenbach remains true to the inductive method.

> If we were to analyze the discoveries of [. . . scientists], we would find that their way of proceeding corresponds in a surprisingly high degree to the rules of the principle of induction [. . .]. The mysticism

of scientific discovery is nothing but a superstructure of images and wishes; the supporting structure below is determined by the inductive principle. [. . . It] seems to be a psychological law that discoveries need a kind of mythology [, ...] that sometimes those men will be best in making inductions who believe they posses other guides. [Reichenbach, H. (1938). Experience and Prediction. An Analysis of the Foundation and Structure of Knowledge. Chicago, Illinois: University of Chicago Press, 403.]

So what have the neopositivists taught us? First of all, induction. Unfortunately, they understood it mainly in the horizons of logic. Carnap and Reichenbach had the opportunity to connect the content of the inductive method with the achievements of mathematical statistics. Nevertheless, they did not.

Positivists love to talk about experimental data, but they say little about the adduction method to get it.

The main mistake of the positivists is the lack of confidence in the concept of theory. As a result, they misunderstand the meaning of the principle of theoretical relativity. In this connection, Mach's mistake is indicative, when he declared to the surprise of many that bodies are complexes of sensations. It was correct to say that both bodies and complexes of sensations are representations of theories. Nevertheless, these representations are different. Therefore, it is inappropriate to consider bodies as complexes of sensations. It is funny that many authors, including Vladimir Lenin from Russia, who did not explain the nature of the objects in any way, nevertheless, were convinced that Mach had made the grossest mistake. He was certainly wrong, but not as crudely as his critics, who misunderstood the nature of objects even more than Mach did.

Another mistake, this time of the neopositivists, was a misunderstanding of the nature of logic. They wanted to achieve clarity in understanding the nature of science, using as a tool one of the sciences, moreover, poorly understood. This way of philosophizing could not lead to the desired success.

In the era of modernity, positivists stood at the origins of the modernity philosophy of science. For historical reasons, representatives of other schools in the philosophy of science have chosen positivists as a favourite target of criticism, often forgetting to acknowledge their achievements. They consisted primarily of the assertion, not sequential, of intratheoretical conceptual transduction with an emphasis on the method of induction.

15. What have critical rationalists taught us?

The founder of critical rationalism is Karl Popper. Imre Lakatos and Hans Albert are the most prominent representatives of critical rationalism. In my opinion, critical rationalism is hard to imagine without the philosophy of Immanuel Kant and Albert Einstein. Popper sometimes referred to the works of Kant and Einstein, but in a way as not to question his importance as the founder of critical rationalism did.

Unlike positivists, critical rationalists put the concept of the theory above all else.

> The empirical sciences are systems of theories. The logic of scientific knowledge can therefore be described as a theory of theories. Scientific theories are universal statements. ... Theories are nets cast to catch what we call 'the world': to rationalize, to explain, and to master it. We endeavour to make the mesh ever finer and finer. [Popper, K. (1959) The Logic of Scientific Discovery. London and New York: Routledge, 37-38.]

> It is the theory, which decides what can be observed. [Einstein, A. (1925) Quoted from Heisenberg, W. (1971) Physics and Beyond. New York: Harper, 63.]

> It can scarcely be denied that the supreme goal of all theory is to make the irreducible basic elements as simple and as few as possible without having to surrender the adequate representation of a single datum of experience. [Einstein, A. (1934). On the Method of Theoretical Physics. Philosophy of Science 1(2): 165.]

In short, the position of critical rationalists is that researches should invent and criticize theories tirelessly. Theories themselves act primarily as the realization of the potential of principles, of what Einstein called irreducible basic elements. Immanuel Kant was perhaps the first philosopher of science to understand the primary importance of principles. He explained the nature of arithmetic, geometry, ethics, and aesthetics according to the principles of time, space, categorical imperative, and taste. Like Kant, Einstein was also unusually purposeful in his search for principles. In this regard, it is significant, in particular, that in understanding the experimental data on the independence of the speed of light in a vacuum from the speed of its source, he thought differently than his colleagues. They wanted to explain them by certain laws. Einstein summarized their meaning as a principle of short-range action. He thought precisely as a critical rationalist.

Popper paid little attention to principles. He, as noted above, understood theory as a set of universal laws. For me, it remains a big mystery why Popper did not pay attention to the importance of principles.

A critical-rationalistic project of the philosophy of science with an emphasis on the concept of theory presupposes particularly close attention towards it. In my opinion, in this respect, Popper surpassed both Kant and Einstein. Popper's predecessors related criticism only to underdeveloped knowledge. He gave it the significance of a method for the growth of all knowledge, and above all, the most developed.

In offering a program of criticism of theories, Popper found himself in an extremely difficult situation. He proposed a certain process of criticism without specifying the method of its implementation. His recipe, to boldly pose problems, set up decisive experiments, and invent new theories, is still not the sought method. There is no answer to the questions: "How to look for new problems?", "How to set up decisive experiments?" and "How to invent new theories?"

Popper's decisive mistake was his underestimation of the relevance of the methods of intratheoretical transduction, especially induction and abduction. They provide both criticism and development of the theory. If a researcher consistently implements the methods of deduction, adduction, induction, and abduction, then he will certainly clarify both the problematic aspects of cognition and the ways of its improvement. Einstein and Popper did not pay attention to this circumstance. They both made intuition responsible for inventing a new theory but did not clarify its content. That is why the recognition of intuition as a method of knowledge turns out to be untenable.

Albert Einstein possessed the rare property of presenting a physical theory in a vivid aphoristic form. However, he was far from always accurate in his expressions. I will cite in confirmation of my words one of the most famous quotes of the great physicist.

> The supreme task of the physicist is to arrive at those universal elementary laws from which the cosmos can be built up by pure deduction. There is no logical path to these laws; only intuition, resting on sympathetic understanding of experience, can reach them. [Einstein, A. (1918) The Collected Papers of Albert Einstein. Princeton, N.J.: Princeton University Press, vol. 7, 44.]

The expression "The supreme task" explains little, I would like more accurately assess the activities of physicists. Einstein stresses the role of universal laws. First, physicists do not know universal laws; the nature of laws is always given only in connection with the data obtained, nothing more. Secondly, the cosmos is not limited to physical phenomena. It is wrong to associate the fate of physics with the cosmos. Thirdly, "pure deduction" is not enough to reproduce physical phenomena, in particular, to explain the results of observations, which involves the implementation of methods of adduction, induction, and abduction. Fourth, it is incorrect to link the achievements of physics with logic. Discoveries are the result of the implementation of physical methods. Logic is to give them a form of calculus. Fifth, the reference to intuition, even when it comes close to the "sympathetic understanding of experience", does not explain anything. In my opinion, "understanding of experience" is provided by induction.

Often, researchers, when discussing the nature of scientific discoveries, adhere to the following extreme. Making sure that the discovery is not described by the operations of some chosen logical theory, they immediately refer it to accomplishment to the realm of intuition, which is extolled as the true instance of creativity. The methods of creativity are deduction, adduction, induction, and abduction. Each of these methods is not easy to implement and involves many search operations. They are often considered to be creative, but they organically express the content of the considered methods, and not something additional to them, i.e. special methods of creativity.

The critical-rationalistic project involves revealing the fate of both a new and partially outdated theory. Albert Einstein has repeatedly expressed himself quite correctly on this score.

> The new theory shows the merits as well as the limitations of the old theory and allows us to regain our old concepts from a higher level. [Einstein, A. and Infeld, L. (1950) [1938] The Evolution of Physics. London: The Scientific Book Club, 158.]

An obsolete theory is a limiting case of a more advanced theory. It also means that the developed theory and all its limiting cases have unity. Einstein came close to understanding the need for the concept of league-theory. In my opinion, Karl Popper and Imre Lakatos less than Einstein were convincing in expressing the unity of an obsolete and new theory. Popper limited himself to stating the existence of more and less developed theories. How exactly are their relations to each other, he did not explain. Lakatos drew attention to the stability of theories; it is far from immediately that a theory gives way to its rival. In my opinion, each cycle of cognition leads to a new theory. Significant differences, however, often do not come immediately.

So what has critical rationalism taught us? First, to work with theories (Popper). Second, to work with the principles of theories (Kant and Einstein). Third, to order theories (Einstein). Fourth, to strive to improve theories (Popper).

16. What have analytic philosophers taught us?

The answer to the above question seems to be obvious. Analytical philosophers Russell and Frege, Moore and Wittgenstein, Carnap and Quine, Austin and Searle, taught us the logical analysis of language. In reality, however, the significance of analytic philosophy goes far beyond the domain of logic and language. First, it refers to the development of the doctrine of the conceptual and methodological structure of science. Starting with a logical analysis of language, analysts were able to present in a new light the entire philosophy of science with its numerous branches.

Russell and Frege, being excellent mathematicians and logicians, undertook grandiose attempts to free arithmetic (Frege) and set theory (Russell) from paradoxes. The means of liberation was logic, above all, the logic of first-order predicates. Logic is the key to understanding the logical form of mathematics. However, Russell and Frege believed that logic is the key to understanding the nature of all sides of mathematics. This was an exaggeration. When defining the nature of mathematics, researchers should take into account the achievements not only of logic but also of all other dependent sciences, in particular, philosophy of science, linguistics, computer science and pedagogy.

Of course, analytic philosophers in one way or another had to turn to the philosophy of science, not limiting it to the philosophy of mathematics and logic. In this regard, Ludwig Wittgenstein ended his "Tractatus Logico-Philosophicus" with the following conclusion.

> The right method of philosophy would be this. To say nothing except what can be said, i.e. the propositions of natural science, i.e. something that has nothing to do with philosophy: and then always, when someone else wished to say something metaphysical, to demonstrate to him that he had given no meaning to certain signs in his propositions. This method would be unsatisfying to the other – he would not have the feeling that we were teaching him philosophy – but it would be the only strictly correct method. [Wittgenstein, L. (1921) Tractatus Logico-Philosophicus, paragraph 6.53.]

Wittgenstein connected the fate of analytical philosophy with the nature of the scientific method. Philosophers do not invent it but distinguish from the thickness of the natural sciences. The nature of the axiological sciences was beyond his attention.

In light of paragraph 6.53, each well-formed sentence has a well-defined meaning. We are talking about logical atomism as opposed to holism. According to Russell, just as matter consists of atoms, the text consists of the last, elementary units of its analysis, separate sentences, i.e. logical atoms. Empirical facts correspond to logical atoms. The appeal of philosophers to language is not accidental. It is much more clearly divided into conceptual units than mentality.

Among the founders of analytical philosophy was George Edward Moore. His work is very indicative of understanding the fate of analytical philosophy. Being a professional ethicist, he, carried away by the logical analysis of language, used it to clarify the content of ethics. Moore concluded that ethical knowledge rests on the ability to understand intuitively fundamental ethical truths for which we can give no basis. With this conclusion, Moore, of course, completely remained in the bosom of metaphysics.

Prominent analytical philosophers have achieved excellent results when they analyzed scientific theories. Such are the studies of Frege in the field of arithmetic, Russell in the field of set theory, Reichenbach in the field of physics. If they, armed with the logical analysis of language, turned like Moore to metaphysical systems, they never reached the desired scientific heights. The main achievement of analysts was not a logical analysis of language, but a logical analysis of the conceptual structure of scientific theories. In this regard, analytical philosophers tend to prioritize scientific methods.

In their works, it is extremely rare to find critical remarks about such methods of intratheoretical transduction as deduction, induction, and abduction. At the same time, they certainly emphasize the actuality of the experimental stage of research and elucidation of the mechanisms of causal relationships. Most analysts, following Willard Quine, consider the experimental stage of research to be universal for all sciences. In this regard, Quine's denial of analytical judgments, supposedly independent of experiments, met with wide support. It means that even sentences of logic and mathematics are tested experimentally insofar as they are part of theories with obvious experimental content.

Analytical philosophers tend to be adept at using intratheoretical transduction techniques. In my opinion, this skill of theirs does not apply to the methods of intertheoretical transduction. In this regard, the positions of Willard Quine and Thomas Kuhn about intertheoretical relations are indicative.

Quine believed that

> … [O]ur statements about the external world face the tribunal of sense experience not individually but only as a corporate body. … Any statement can be held true come what may, if we make drastic enough adjustments elsewhere in the system. [Quine, W.V. (1980) From a Logical Point of View. Cambridge Massachusetts: Harvard University Press, 41, 43.]

From these positions, there is no need to build league-theories. As a set of statements, the same theory, depending on the assumptions made in world science as a whole, can be considered the most developed and less developed, and unchanged. In my opinion, Quine did not take into account that the progress of science is accomplished through the multiplication of cycles of conceptual transduction. Their connection with each other is not arbitrary, it is such that the tapestry of world science is given in a quite definite way, including as a well-known correlation between less and more developed theories.

The analysis of inter-theoretical relations occupies a prominent place in the work of Thomas Kuhn. His key claim is that any two theories are incommensurable with each other.

> To state that two theories are incommensurable means that there is no neutral language, or other type of language, into which both theories, conceived as sets of statements, can be translated without remainder or loss. [Kuhn, T.S. (2000) The Road Since Structure: Philosophical Essays, 1970-1993, with an Autobiographical Interview. Chicago and London: Chicago University Press, 36.]

In my opinion, the weakness of Kuhn's position is that he does not consider the methods that lead to new theories. Any cycle of intertheoretical transduction leads to the emergence of a new theory. There is no need to involve the concept of neutral language in the analysis. Theories themselves are incarnations of languages.

They are commensurate with each other in the sense that a less developed theory is the limiting case of a more developed theory.

The vast majority of analytical philosophers gravitate towards the language and behavioural representations of the theory. Many of them avoid talking about values. Otherwise, they replace the behavioural view with a pragmatic one.

Today analytic philosophers are our main teachers in understanding the philosophy of science. In many ways, they adopted the achievements of the positivists and critical rationalists. However, not all analytic philosophers succeed in developing scientific work successfully. Among analytic philosophers, metaphysical tendencies are far from being eradicated. As a rule, they make themselves felt when the nature of some phenomena is considered regardless of the data of the sciences. The detachment of science from philosophy is a direct path to metaphysics.

17. What have the Marxists taught us?

The previous three essays dealt with distinctly science-based philosophical direction, namely, the 'great trio' of positivism, critical rationalism, and analytic philosophy. Representatives of the 'great trio' give the most balanced characterization of the conceptual and methodological structure of the sciences. Now is the time to turn to those philosophical directions that are scientifically inferior to the 'great trio', but also possess several achievements. Let us turn in this connection, first, to Marxism, the brainchild of Karl Marx and his colleague Friedrich Engels.

In 1848, they published the famous "The Communist Manifesto". Marx and Engels made an impassioned appeal to the human community for economic and social justice. They believed that this appeal expresses above all the aspirations of the working class. Until 1859, Marx did not possess any coherent conceptual and methodological theory. He worked it out in his book "Contribution to the Critique of Political Economy". In 1867, Marx published the first volume of "The Capital". Its conceptual and methodological structure is the same as that of «Contribution to the Critique of Political Economy.»

Marx proceeded from the dialectical method of Georg Hegel, in contrast to him, arguing that the consciousness of people is secondary because production (economic) relations determine it. Following Hegel, Marx argued that the dialectical method is an ascent from the abstract to the concrete. With this statement, if we understand it as a reproduction of a system based on its elements, perhaps a modern scientist will agree. The main peculiarity of Marx's method lies in something else, namely, in its saturation with contradictions. I tried to find a quote from Marx that would express his innermost understanding of the scientific method. Nevertheless, I never found it. Friedrich Engels' review of K. Marx's "Contribution to the Critique of Political Economy" explains a lot. He wrote the article with the knowledge of K. Marx and it contains the following passage.

> [W]e begin with the first and simplest relation which is historically, actually available, thus in this context with the first economic relation to be found. We analyse this relation. The fact that it is a *relation* already implies that it has two aspects which are *related to each other*. Each of these aspects is examined separately; this reveals the nature of their mutual behaviour, their reciprocal action. Contradictions will emerge demanding a solution. But since we are not examining an abstract mental process that takes place solely in our mind, but an actual event which really took place at some time or other, or which is still taking place, these contradictions will have arisen in practice and have probably been solved. We shall trace the mode of this solution and find that it has been effected by establishing a new relation, whose two contradictory aspects we shall then have to set forth, and so on. [Engels, F. (1859) Karl Marx: In Marx, K. (1970) [1859] Contribution to the Critique of Political Economy. Edited by M. Dobb. Moscow: Progress Publishers, 225-226.]

According to Marx and Engels, the study of relations is associated with the need to ascend from one contradiction to another. The ascent from the abstract to the concrete is at the same time an ascent from undeveloped contradictions to more developed ones. This conclusion is the root cause of the many misadventures of Marxism. A contradiction is the incompatibility of two concepts or statements. It is contradictory to assert the truth of two statements: "Peter is Mary's brother" and "Peter is not Mary's brother." If one relationship is fixed, "Peter is Mary's brother," then there is no contradiction. The scientific method does not consist of the transition from one contradiction to another; it consists of the sequential implementation of deduction, adduction, induction, and abduction. If there are contradictions in scientific research, then they indicate that it has gone astray. Both Marx and Engels were well aware of the content of deduction and induction; nevertheless, they associated the content of the scientific method with the ascent from undeveloped contradictions to more developed ones.

Thus, Marx and Engels did not possess a genuinely scientific method. Of course, this circumstance could not pass without leaving a trace for the quality of their scientific research. This also applies to Marx's most famous work, the three-volume "The Capital". Its fundamental position states that the labour contained in goods is dual it represents the unity of concrete and abstract labour. Marx was most proud of the fact that he discovered of abstract labour. Meanwhile, the concept of abstract labour is scientifically unsound.

Marx believed that two varieties of concrete labour, for example, the labour of a weaver and a locksmith, have not only different but also the same quality, which is evidenced by a single monetary quantitative measure for them. This quality is inherent in abstract labour. There is no need to drive the concept of abstract labour. Economists evaluate the work of people uniformly, namely, in monetary units, by the principles of economics. This circumstance determines the economic quality of labour. Economic labour, not concrete and abstract labour in their unity, produce the goods.

Without knowing the scientific method, Marx, naturally, could not determine correctly the future of humankind. His prediction of the triumph of first socialism and then communism turned out to be a dubious metaphysical prediction. The collapse of the USSR, which happened in 1991, was not a triumph of science, but evidence of the perniciousness of attempts to replace the philosophy of science with dialectical metaphysics. Fortunately, such a substitution could not take place in full. There was real science in the USSR, not thanks to, but in many respects contrary to the Marxist methodology of science.

There is no reason to believe that Marx and Engels underestimated the relevance of science. They quite sincerely sought to use its achievements for the benefit of humanity. Moreover, both purposefully studied the nature of the sciences. There was even a division of labour between them. Marx studied the nature of the social sciences, and Engels studied the nature of the natural sciences. However, both researchers have failed to develop a consistent conception of the philosophy of science. The fact is that they looked at sciences through the eyes of dialecticians and, as a result, saw in them precisely its triumph. The main lesson in the work of Marx and Engels is the need for a thorough elaboration of the philosophy of science. In its absence, even the most beneficial appeals, such as social justice, invariably lead to disaster.

The failure of the Marxists does not turn us away from the principle of justice. It shows that good goals need their correct conceptual and methodological support.

18. What have the supporters of pragmatism taught us?

In the last quarter of the 19th century, Charles Sanders Peirce and William James established pragmatism as a philosophical trend. Somewhat later, John Dewey proved himself as an outstanding pragmatist. Peirce formulated the maxim of pragmatism.

> Consider what effects, that might conceivably have practical bearings, we conceive the object of our conception to have. Then, our conception of these effects is the whole of our conception of the object. [Pierce, C.S. 1878 How to Make Our Ideas Clear. Popular Science Monthly 12(1), 293.0

If ideas remain closed in the sphere of mentality, as does Locke, Berkeley, and Hume inherit the case within the framework of Cartesianism from Descartes, then they are confused and unreliable. Theories are strong in their practical implications as tools for solving vital problems.

When evaluating pragmatism, it is extremely important to keep in mind that its founders had advanced ideas about the scientific method for their time. Peirce is the author of the methodological triad abduction – deduction – induction. He knew very well that in scientific research these methods form an inseparable whole. Dewey, in his book "How We Think", competently wrote about scientific thinking as a sequential succession of induction and deduction. Dewey considered the experiment to be the main resource of scientific thinking because it helps to highlight significant elements of a rough, vague whole. Unlike the Marxists, pragmatists were supporters of scientific methods, not only in words but also in deeds.

I mentioned the scientific awareness of pragmatists for a reason. It should shed some light on their choice of pragmatic orientation. It consisted, first, in the emphasis on experimentation. Second, they took into account that experiments may be far from the demands of life. Consequently, scientific thinking should contribute to the resolution of vital, i.e. practical problems.

At first glance, the philosophical choice of pragmatists seems to be a completely consistent rational measure. This impression needs correction in the light of the principle of theoretical representation. According to this principle, any problems are representations of theories. Scientific theories are not instrumental. Outside of the science are not problems. Problems are the result of the cultivation of theories, and their solution appears as the need for their improvement.

Another weakness of pragmatism consists of exaggerating the role of the experiment and, therefore, underestimating the importance of the other three stages of intratheoretical transduction. As a result, the experiment is opposed to predictions, data processing, and updating the principles of the theory. Thus, pragmatists in the cultivation of intratheoretical transduction are not consistent enough. However, they remain within the bounds of a scientifically oriented philosophy. It is not surprising that pragmatism, despite its age, more than a century remains a vital program. Of course, the development of pragmatism met with significant difficulties. In particular, this concerns his relationship with analytical philosophy, with which he has to coexist in the American open spaces.

The fact is that the representatives of analytical philosophy developed a standard for the philosophy of science, which was not in traditional pragmatism. The pragmatists were mostly strong in the axiological sciences. Analysts shone in the fields of logic, mathematics, and physics. The impression was that pragmatists were not destined to master the achievements of analytical philosophy. Help came from analytical philosophers. The expansion of the area of analytical philosophy needs generalizing ideas. In this regard, pragmatic ideas, with their emphasis on practice and current democratic transformations, came in very handy. Richard Rorty, Hilary Putnam, Donald Davidson, and Willard Quine renewed pragmatism. Neo-pragmatism has replaced traditional pragmatism. Proponents of neo-pragmatism usually reject the concept of universal truth, realism, an objectivist view of society, the opposition of facts and values, but support pluralistic views of the world, democratic processes, and the use of language games in practice.

The classical pragmatism has posed many problems comprehension of which needs a high-level philosophy of science. In this regard, quite rightly high hopes are pinned on the representatives of analytical philosophy. The enrichment of classical pragmatic projects with the achievements of analytical philosophy is important for the development of special philosophy of science, for example, philosophy of technology, philosophy of history, and philosophy of economics.

What have the pragmatists taught us the most? They taught us an active life position, combined with experimental enthusiasm. What is the main disadvantage of pragmatism? In the exaggeration of practice, i.e. of the practical representation of theories.

I believe that the orientation of pragmatism became the main feature of North American civilization. Its future is an open question.

19. What have phenomenologists taught us?

Phenomenology is one of the modern philosophical trends. Many philosophers have sought to implement a phenomenological project, which is that the fundamental philosophical provisions are the result of the purposeful reflexive work of the researcher, carrying out the synthesis of various perceptions. In this regard, the studies of the German philosopher Edmund Husserl are the most exemplary. His work is undoubted of considerable interest.

Husserl is concerned about the state of affairs in both philosophy and science. Since antiquity, philosophy was exemplary knowledge. Its influence extends to all sciences. The authenticity of philosophy determines the authenticity of science. In this regard, it is necessary to decide, first, with philosophy. René Descartes and Immanuel Kant largely determined the specifics of philosophy. Descartes, like no one else, emphasized the paramount importance of human subjectivity in philosophizing. The belittling of subjectivity invariably leads to the emasculation of the vitality of philosophy, and after it, of all science. Kant, with his transcendental schematics, expressed the importance of philosophy for the sciences. Nevertheless, he failed to give philosophy a phenomenological form. Husserl takes on the difficult task of creating phenomenological transcendence.

Not only philosophy but also science is constantly in Husserl's field of vision. The state of affairs in the latter reinforces his desire to create a full-fledged phenomenology. He believes that the ideal of the physical and mathematical knowledge of Euclid-Galileo prevails in modern science. Euclid developed a highly extraordinary axiomatic method. The specificity of Euclidean geometry lies in the fact that scientists operate with ideal forms as limiting cases of ontological certainties. Within the framework of Galilean physics, scientists united the world of Euclidean idealizations with the world of bodily things and empirical dimensions. Mathematics becomes the language of physics, and later of technology. The ideals of objectivism, naturalism, formalism, and mechanization triumph in science. The subjectivity and intersubjectivity of people fell into the shadow of these ideals. This testifies to the oblivion of the life world of people, in other words, the crisis of the sciences designed to ensure the well-being of people.

Thus, starting from perceptions, Husserl strives to generate the entire complex conceptual world of philosophy and science, avoiding the ravines of not only objectivism and naturalism but also subjectivism and relativism. He is an adherent of true knowledge, and it is incompatible with subjectivism and relativism.

Initially, Husserl was optimistic for his psychological training. He meant that perceptions as elements of synthesis belong to psychology. However, he quickly realized his mistake, becoming one of the adamant critics of attributing philosophical functions to psychology. It remained unclear why philosophy begins precisely with perceptions. If Husserl considered philosophy to be one of the sciences, he would be able to explain its status, including the ability to work with perceptions. However, like all supporters of transcendentalism, Husserl regarded philosophy as the first science. In this case, its nature is given, but not explained. The researches can absolutize the meaning of not only logic and psychology but also philosophy.

To Husserl's credit, he describes the transition from perceptions to philosophical concepts as an extremely multi-tiered process. However, it is very significant that Husserl invariably attaches decisive importance to variations of perceptions and the discernment of their essence through intuition. Husserl believed that this was the only way to ensure the authenticity of philosophy. Numerous proponents of phenomenology adhere to the same position. Their phenomenological enthusiasm tends to be impressive. However, I tend to doubt its productivity.

I am doing this due to the following circumstance. According to the plan, the phenomenological project was to lead to a significant transformation of the sciences. Did it take place? If so, then it testifies in favour of phenomenology. Otherwise, the phenomenological project must be called into question.

None known to me realizations of the phenomenological project outside of philosophy, in particular, in mathematics (Husserl), sociology (Alfred Schütz), economics (Till Düppe), ends with any significant restructuring of these sciences. The phenomenologists offer the conceptual filling of sciences in a mental form, but it remains the same in content. They use the same principles, laws, and variables as other scientists. How can this circumstance be explained? From the standpoint of the theory of conceptual transduction, this can be done quite simply.

As you know, each theory has many representations, including language and mental. The mental representation of a theory can be cogitative or sensory-emotional. Phenomenologists emphasize the sensory-emotional representations of theories. They absolutize these representations, giving it the meaning of the human lifeworld. This means that the phenomenological project of representing science complements other projects, rather than refutes them, as Husserl believed. The phenomenological project is not the only correct one. Nevertheless, and it is relevant. Without it, science looks impoverished.

Does philosophy have a transcendental character? Of course not. Philosophy of science is one of the sciences, not the first of them. Its purpose is that it summarizes the conceptual and methodological achievements of the sciences and supplies sciences with this product. Phenomenological philosophy can be successful only when it presents these achievements in a sensory-emotional form. In this regard, I again draw the reader's attention to the relevance of the methods of intratheoretical and intertheoretical conceptual transduction. Husserl and his supporters passed them by. Contrary to their opinion, science does not see the essence of things through intuition in variations of perceptions. The methods of conceptual transduction, not intuition provide true knowledge of the nature of things,

So, what have phenomenologists taught us, not just in words? To use sensory-emotional representations of theories. It remains only to regret that on this path they failed to avoid significant conceptual and methodological errors.

20. What have the hermeneutists taught us?

Hermeneutics, like phenomenology, is an influential philosophical trend of our time. Most of all, hermeneutists are concerned with the issue of mutual understanding between people. It was in the centre of attention of the hermeneutics of three trends, the hermeneutics of consciousness (Rudolf Schleiermacher and Wilhelm Dilthey), the hermeneutics of being (Martin Heidegger and Hans-Georg Gadamer), and critical hermeneutics (Karl-Otto Apel and Jürgen Habermas).

As a rule, hermeneutists pay considerable attention to finding a method that is adequate for the specifics of axiological sciences.

On this score, the position of the German historian Johann Droysen has become widely known.

> According to the objects and according to the nature of human thought are the three possible scientific methods: the (philosophically or theologically) speculative, the mathematical-physical, and the historical.
>
> Its essence is to recognize, to explain, and to understand. [Droysen, J.G. (1868). Grundriss der Historik [Floor Plan of the History.]. Leipzig: von Veit & Comp, 11. (In German).]

The concern of the hermeneutists is understandable. Each person is different. How do people agree, without which they can hardly manage?

Friedrich Schleiermacher clarified the main provisions of hermeneutics by examining the activities of a translator. First, he cannot avoid considering the text from a linguistic standpoint. Secondly, he must understand the individual psychological characteristics of the author expressed in the text. Thirdly, it should be taken into account that the position of both the author and the translator is conditioned by a certain socio-cultural context. The constant movement between the socio-cultural context (the whole) and the individual characteristics of the author (the partial) forms a hermeneutic circle. A translator achieves complete success only when he understands the author better than he understands himself.

Wilhelm Dilthey turned the entire hermeneutic program towards science. The following features are characteristic of the natural sciences: causal explanation, the unity of processes function with general laws, lack of a personal aspect. For the human sciences are characteristic understanding, a specific context, constant transformations of the individual and society. The methodology of the human sciences should provide an understanding of the life experience of people and its historical transformations.

Schleiermacher and Dilthey, expressing the conceptual content of hermeneutics, encountered significant difficulties. They tried to realize the union of philosophy and psychology but failed to implement it successfully. Moreover, this despite their philosophical preferences, which were Hegel's dialectics for Schleiermacher and Bergson's philosophy of life for Dilthey.

Martin Heidegger gave the hermeneutics a new twist. He understood it not as a methodology of cognition, including of science, but as a method of the genuine, i.e. creative human existence. Only as a hermeneutic being, successfully destructuring the established conceptual canons is a person able to be-in-the-world. Language is a decisive instance of being. Accordingly, in the process of understanding, its language form prevails always boiling down to finding the exact word.

Starting from the ideas of Heidegger, Hans-Georg Gadamer presents the hermeneutics of being as a kind of systemic phenomenon. Unlike Heidegger, he collects under the heading 'hermeneutics' all the most valuable of philosophy known to him. In my opinion, the result is such a chain of conceptual nodes: tradition, understood as a language phenomenon and the inclusion of people in its history – the dialectic of questions and answers – dialogue, a hermeneutic circle between the part and the whole – the symbolization of the whole by an image – art as the scale of philosophy – building up hermeneutic experience – expanding hermeneutic horizons – understanding as an intersection of hermeneutic horizons – unity of word and deed – ensuring the triumph of a common cause.

The Heidegger-Gadamer hermeneutics of being is subject to significant criticism from Karl-Otto Apel and Jürgen Habermas. According to Apel, contemporary hermeneutics should be transcendental, pragmatic, communicative, and reflective. All these features are present in the hermeneutics of being, but not in a sufficiently developed form. Apel distinguished between scientific and hermeneutic rationality, which complement each other. The scientific explanation operates with the causes of actions, and the hermeneutic with the intentions of people. In both cases, there is a certain combination of explanation and understanding. Researchers who are not devoid of intentions carry out the scientific explanation. In the composition of hermeneutic rationality, understanding is in the first place, but it involves the use of explanations. Apel believed he had succeeded in combining American pragmatism and analytical philosophy with German transcendentalism and hermeneutics.

Jürgen Habermas's hermeneutic position is in many ways reminiscent of Apel's transcendental hermeneutics. However, unlike Apel, he is especially concerned about the fate of the social sciences, primarily sociology. Habermas rethinks the historical development of the social sciences, including Marxism. In this regard, one could expect from him a significant renewal of the understanding of the relationship between hermeneutics and social sciences. These expectations did not come true. Like his predecessors in hermeneutics, he sees it as the key to the reconstruction of all social sciences.

Habermas pays special attention to the ways of reaching an agreement, the main content of mutual understanding between people. Each participant in the discourse initially can understand the meanings of norms, both their own and those of others. If participants in the discourse are sincere in presenting their values, do not exclude anyone from discourse, and reason according to the laws of logic then they inevitably agree on common values. Thus, hermeneutics leads to an ethics of responsibility. It, in turn, is a condition for the reconstruction of the social sciences.

So what have the hermeneutists taught us? First, to deal with the originality of the human sciences in their comparison with the sciences of nature. Secondly, to seek the relevance of communication between people. Their long-term efforts indicate that the problems they have identified are not resolved by turning to philosophy as a metaphysical premise of the sciences. The correct line of reasoning is that the natural, axiological and formal sciences are initially considered. Then their achievements are fixed as conclusions of the philosophy of science. After that, it turns out that the methods of various sciences are similar to each other. Scientifically sound conclusions should be built by the methods of conceptual transduction. Only then is it found that mutual understanding presupposes the intersection of theories. Mathematicians, physicists, economists, and philosophers understand each other only when their theories have common, intersecting fragments.

21. What have poststructuralists taught us?

Poststructuralists are philosophers who deny any established philosophical canons, in particular, systems, structures, identities, concepts. They focus on the mutability of phenomena, the freedom of individuals, pluralism. Michel Foucault, Gilles Deleuze, Jacques Derrida, and Jean-François Lyotard were prominent poststructuralists.

The meaning of the term 'poststructuralism' is not clear. The group of philosophers in question denied not only structuralism but also many other philosophical projects. In my opinion, the term 'differism' is well suited to denote the philosophical trend under consideration. Deleuze is the author of the book "Différence et Répétition". For Deleuze, différence is the transcendental principle, the condition for the appearance of empirical diversity. Derrida developed the theory of deconstruction. In it, the most relevant concept is différance, a term coined by Derrida to denote difference and deferral of meaning. Lyotard's book, "Le Différend", is the best summary of his philosophy. As we can see, Deleuze, Derrida, and Lyotard give central meaning to terms with a common part differ. It is in this connection that I propose the term «differism.» Let me turn to a more detailed characterization of the views of prominent poststructuralists.

The fundamental concept of Michel Foucault's philosophy is discursive practices as a set of statements, discourses. Everything that exists, including people and things, are manifestations of discursive practices. Their efficient production requires some methodologies. According to Foucault, this is genealogy, criticism, and problematization. Genealogy means that discursive practices are considered in the historical field, in some specific conditions. Criticism removes all kinds of prohibitions that hinder the freedom of discourse. As a rule, these prohibitions come from some authorities. Problematization gives the ratio of true and false a playful, creative character. Foucault characterizes the all-round development of discursive practices as the will to truth. Truth is opposed to the undeveloped, enslaved, and unfree. The will to truth shapes an ethic of self-care that is grounded in the daily existence of people and the aesthetics of the body.

We must pay tribute to Foucault, in my opinion; he quite skillfully weaves interdisciplinary connections into his philosophizing. Foucault does this by no means accidentally. Any science delegates part of its powers to auxiliary sciences, in particular, law, the science of power, ethics, and aesthetics. We see how topics related to these sciences now and then come up in Foucault's analyzes. He makes every effort to present the topics he is considering, whether it be bio-power, the formation of the penitentiary system, and issues of sexuality in the most complete form.

Unfortunately, the methodological part of Foucault's philosophy, primarily concepts of genealogy, criticism, and problematization, raises great doubts. In science, these concepts develop within the framework of conceptual transduction. Foucault avoids this path. As a result, he remains outside the philosophy of science. The characteristics of genealogy, criticism, and problematization turn out to be very vague and imprecise. He fails to find an alternative to the philosophy of science. Ultimately, Foucault remains within the framework of metaphysics.

It is interesting that Jacques Derrida, another outstanding differist, quite deliberately seeks to transcend metaphysics. He once explained his position as follows.

> At the point at which the concept of différance and the chain attached to it, intervenes, all the conceptual oppositions of metaphysics (signifier/signified; sensible/intelligible; writing/speech; passivity/activity; etc.)- to the extent that they ultimately refer to the presence of something present (for example, in the form of the identity of the subject who is present for all his operations, present beneath every accident or event, self-present in its "living speech," in its enunciations, in the present objects and acts of its language, etc.) – become non pertinent. They all amount, at one moment or another, to a subordination of the movement of différance in favor of the presence of a value or a meaning supposedly antecedent to différance, more original than it, exceeding and governing it in the last analysis. This is still the presence of what we called above the "transcendental signified". [Derrida, J. (1981) Positions. Trans. by A. Bass. Chicago: University of Chicago Press, 30.]

Derrida's way of philosophizing is eminently eccentric. Dissatisfied with all kinds of centrism, he begins his philosophizing with metaphysical oppositions. Derrida turns oppositions into aporias and continuously transforms them. Philosophizing turns into an aporetic transgression without any rules. Everything stable, including things, concepts, and signs, turns into grams. Derrida did not accept any methods. Nevertheless, he often used the following techniques. First, he formulates binary oppositions (speech/writing, answer/silence, etc.). Second, Derrida gives them the character of aporias. Third, he goes from one aporia to another. Fourth, Derrida dissects the words, implements morphological analysis and uses dictionary entries and translations of terms from one language to another. Fifth, he gives any rule or concept a recurrent character, A turns into not-A.

Most analytical philosophers strongly oppose the philosophy of Derrida. They accuse him of relativism and rejection of the canons of science. Neither Derrida himself nor the supporters of his doctrine have revealed the flaws of science. They are not proved but are assumed. Derrida proposed, in fact, a program of deconstruction of modern science. This program is rarely taken seriously by any of the experts in science.

Jean-François Lyotard is perhaps most famous for the book "The Postmodern Condition: A Report on Knowledge" (1979). In the most complete form, he outlined his conception in the book "The Differend: Phrases in Dispute" (1983). Lyotard distinguished three types of knowledge, narrative, scientific, and postmodern. Their central concepts are, respectively, the authority of the author of the narration, the truth achieved through proof and practical effectiveness. The state of postmodernity has challenged scientific knowledge. The agonistics of language games prevails; everyone pursues their own goals. Positions of participants in the discourse are incommensurable with each other, so people cannot agree. The discourse is radicalized so much that it turns into strife, le different.

> I would like to call a differend the case where the plaintiff is divested of the means to argue and becomes for that reason a victim. If the addressor, the addressee, and the sense of the testimony

are neutralized, everything takes place as if there were no damages. A case of differend between two parties takes place when the regulation of the conflict that opposes them is done in the idiom of one of the parties while the wrong suffered by the other is not signified in that idiom. [Lyotard, J.-F. (1988). The Differend: Phrases in Dispute. Minneapolis: University of Minnesota Press, 9.]

Like Foucault and Derrida, Lyotard did not prove himself to be an expert in scientific knowledge. His criticism of this knowledge is very superficial. That is why Lyotard's philosophy does not lead beyond the limits of metaphysics.

Poststructuralists believe that postmodern knowledge is superior to science. This superiority manifests itself in a special sensitivity to differences, the ability to appreciate the incommensurable, in the invention of sublime paralogies. By this imaginary superiority, ethics, aesthetics, and politics are being rethought.

Poststructuralists have developed a program for the development of sophisticated individualism and pluralism. In my opinion, it deserves the close attention of scientists.

22. On the diversity and unity of philosophical directions

Because of many years of research, I concluded that science is unthinkable without maximizing diversity (differentiation), accompanied by maximizing the unity (integration) of philosophical trends. The two maximization processes accompany each other. Diversity without unity affirms relativism with its unacceptable refrain "Everything is allowed!" Unity without diversity is implanted with reinforced concrete absolutism: "One step to the left, one step to the right – shooting!" In the philosophy of science, philosophical diversity is especially clearly represented by the coexistence of positivism, critical rationalism, analytical philosophy, phenomenology, hermeneutics, and poststructuralism. The key figures of these philosophies, respectively, Rudolph Carnap, Karl Popper, Willard Quine, Edmund Husserl, Hans-Georg Gadamer, and Jacques Derrida, jointly affirmed philosophical pluralism. At the same time, they have prepared an undoubted surprise for us – none of them is a skillful synthesizer of philosophical trends. This means that in modern philosophy, pluralism dangerously coexists with relativism. As we can see, a problematic situation has developed in the philosophy of science that raises numerous questions.

Each philosophical direction focuses on quite definite concepts. This is for neopositivism – experiment and induction, for critical rationalism – the predictions of the theory and the refuting of the theory, for analytical philosophy – the logical representation of language, for phenomenology – the mental representation of the theory, for hermeneutics – communication for reaching the consensus, for poststructuralism – the originality of the position of each individual. Why do different philosophical directions arise? Because it is impossible to focus on all scientific concepts at once. Significant accents become the embryonic nodes of philosophical trends. Their thickening leads to the isolation of philosophical schools. For example, a hermeneutist unites with a hermeneutist, and not with a critical rationalist or positivist. In modern philosophy, traces of philosophical isolationism are visible everywhere. Nevertheless, philosophical isolationism is not necessary.

How can isolationism be overcome? Being in one direction, to open doors towards other directions, seeking to present them all as parts of the philosophy of science as a whole. Following the neopositivists, one can stubbornly argue that the beginning of a theory is an experiment and inductive processing of its data. At the same time, no prohibitions can be seen for giving credit to critical rationalists, asserting that deductive prediction precedes experiment. Thus, isolationism is not necessary. Why does he triumph in philosophy, displacing the unification of its various directions into the shadows? Apparently because the whole as an object of cognition is much more complicated than its separate part. Reducing the whole to one or even several parts is the foundation of philosophical isolationism. The meaning of the philosophical whole escapes from its representatives. The same is true for pluralism. This time, unconsciously, the parts of the whole are alienated from each other. There is no genuine contact between them.

A variety of philosophical trends is necessary, but it is not enough for meaningful philosophizing. In this regard, the question of the possibility and consistency of their integration acquires fundamental importance. I do not know a single philosopher who would successfully integrate, if not all, then at least the main philosophical trends. This despite the abundance of pluralists and monists in philosophy. Nevertheless, philosophical integration is quite consistent. It is important to understand the way to do it. In this regard, in my opinion, it is necessary to consider the philosophy of existing scientific theories together with the links that hold them together. Then an exciting picture of modern science as a single whole, shimmering with colours of all existing sciences, will open before you. I am inclined to believe that I have gone this way in my books many times. Its methodological basis is the sequential implementation of one after another intratheoretical and intertheoretical transduction. Through the efforts of people, science functions as a self-developing whole.

Why is the integration of philosophical trends the Cinderella of philosophical research? In my opinion, the well-known timidity of philosophers, who strive to limit their research to the study of one or two sciences, is evident. Perhaps they fear the diversity of sciences, the development of which involves a difficult path of acquiring encyclopedic competence. Nevertheless, effective philosophizing is impossible without encyclopedic competence.

Chapter 3.
Special Philosophy of Science, or the Nature of Sciences

23. The nature of physics

Each science has in its arsenal such achievements that radically restructure people's thinking. In this respect, the achievements of physics are very indicative, especially the four revolutions represented, respectively, by classical, relativistic, quantum, and quantum field mechanics. This gait is as follows:

$$T_{CM} \rightarrow T_{RM} \rightarrow T_{QM} \rightarrow T_{QFM}.$$

Mechanics deals with similar features of several types of interactions. For example, quantum mechanics deals with similarities of electromagnetic, weak, and strong interactions. The sciences formed by identifying similar features of the studied phenomena are formal. Mechanics is a formal physical theory. Its separation makes sense only insofar as each type of physical interaction is the subject of some theory.

When considering the nature of science, one should always proceed from the achievements of the most developed theories, which, in contrast to their alternatives, lose sight of a smaller number of factors. In our case, this is the quantum field theory. Fields coexist with particles. There is no field without particles or particles without a field. Particles act as field excitations. The field is much more complex than particles. To describe any particle, it is sufficient to use a finite number of variables (generalized coordinates). In the case of a field, the number of these variables becomes infinitely large. This is not difficult to understand, because, in the case of a field, exhaustive information must be given about the numerous spatial-temporal regions of the field, the number of which cannot be counted. With this in mind, they say that the number of degrees of freedom of a physical field is equal to infinity.

The fundamental concepts of any scientific theory are principles. In the case of quantum field theories, this is, firstly, the principle of stationary action, secondly, the principle of gauge invariance, and thirdly, the principle of short-range action. According to the principle of stationary action, the action of fields and particles takes on an extreme (smallest, largest, or stationary) value. The principle of gauge invariance (symmetry) is that the interactions predicted by the theory do not change if a certain quantity is freely chosen locally. The principle of short-range action is that the speed of any physical signal cannot be greater than a certain value, equal, in particular, to the speed of light in a vacuum. Gauge invariance testifies in favour of the conservation law of some intrinsic property, for example, electric charge. Conservation laws are fulfilled due to the presence of quanta of interactions. Gauge invariance has given the entire quantum field theory unprecedented integrity.

Extreme action, the mechanism of its realization (interaction), symmetry, and conservation laws form an organic unity, which is characteristic just for quantum field theory.

Knowing the nature of quantum field theory is key to understanding the nature of less advanced physical theories. In quantum mechanics, the value of the field is not taken into account; as a result, the ontology of particles dominates in it. Their number of degrees of freedom is finite. There are no creation and annihilation operators of particles in quantum mechanics.

In relativistic mechanics, the concept of a wave function is not used, as a result, the probabilistic nature of physical processes is completely ignored. In classical mechanics, the principle of short-range action is not used. As a result, relativistic effects are not considered. Knowing the nature of quantum field theory is key to understanding the nature of less advanced physical theories. In quantum mechanics, the value of the field is not taken into account, as a result, the ontology of particles dominates in it. Their number of degrees of freedom is finite. There are no creation and annihilation operators of particles in quantum mechanics.

The relativistic theory of gravitation destroys the relatively blissful quantum-field picture of the world. Electromagnetic, weak, and strong interactions cannot be isolated from gravitational interactions, and they do not lend themselves to the already known methods of quantization. This is the main challenge to which physicists have to find an adequate answer.

The rise of physics to quantum field heights has made it extremely difficult to understand its nature. As one of the creators of QFT said somewhat coquettishly, "I think I can safely say that nobody understands quantum mechanics." (Feynman, R. The Character of Physical Law. Cambridge, Massachusetts, MIT Press, 1995, 129). We are talking about the fact that the understanding of modern physical theories is never complete. You cannot put a full stop. It is imperative to continue efforts to master physical theories. This is largely due to the somewhat strange union of physics and mathematics.

When Isaac Newton became the first to write down the law of motion of bodies in differential form, the interpretation of its content caused great difficulties. They increased after quantum mechanics began to represent physical quantities as operators acting on the wave function. However, a matrix form of quantum mechanics is also possible. It turns out that the nature of physical objects is such that more and more advanced mathematics is needed to understand it. Add to this the extremely sophisticated technique and methodology of physical experiments and you will understand that physicists have to solve the most difficult problems, in particular, to identify those objects, for example, particles and fields, with which they have to deal. Moreover, they do it! The discovery of Hicks bosons is extremely revealing in this sense. Their existence was predicted in 1964, but they were identified in experiments only in 2012.

In my opinion, to the above-mentioned union of physics and mathematics, a third party should certainly be added, the philosophy of science. Albert Einstein noted that

> [t]he physicist is compelled by the present difficulties of his science to deal with philosophical problems largely than was the case with earlier generations.

[Einstein, A. (1946) [1944] Remark on Bertrand Russell's Philosophy of Knowledge. (In German). In The Philosophy of Bertrand Russell. Ed. by Schillp, P.A. Evanston, Illinois: Northwestern University, 278.)

In a letter to Robert Thornton, he noted

I fully agree with you about the significance and educational value of methodology as well as history and philosophy of science. So many people today – and even professional scientists – seem to me like somebody who has seen thousands of trees but has never seen a forest. Knowledge of the historic and philosophical background gives that kind of independence from prejudices of his generation from which most scientists are suffering. This independence created by philosophical insight is – in my opinion – the mark of distinction between a mere artisan or specialist and a real seeker after truth. [Einstein to Thornton, 7 December 1944, EA 61–574.]

Einstein is certainly right. Nowadays, from a discerning physicist, not only a narrow specialization is required, but also the broadest possible scientific outlook, the formation of which largely depends on the philosophy of science.

24. Newton's breakthrough to the classical paradigm in physics

Isaac Newton's great achievement was that he gave classical mechanics and with it all physics, the character of a paradigm, of a model. He found exemplary solutions to numerous complex issues, the potential of which has been multiplying for more than three centuries. How did he do it? It is hardly possible to give an unequivocal answer to this question; nevertheless, it is good already because it sets you up for a philosophical mood. The famous book of Newton from 1687 is "Mathematical Principles of Natural Philosophy". Why is he talking about mathematics? Here is Newton's clarification.

[S]eems to be one of the most difficult things that Philosophy needs, I hope to show – as it were, by my example – how valuable mathematics is in natural Philosophy. I, therefore, urge geometers to investigate Nature more rigorously, and those devoted to natural science to learn geometry first. Hence the former shall not entirely spend their time in speculations of no use to human life, nor shall the latter, while working assiduously with a preposterous method, perpetually fail to reach their goal. But truly with the help of philosophizing Geometers and Philosophers who practice Geometry, instead of the conjectures and probabilities that are being marketed everywhere, we shall finally achieve a natural science secured by the highest evidence. [The Optical Papers of Isaac Newton, vol. 1, ed. A. E. Shapiro. Cambridge: Cambridge University Press, 1984, p. 89.]

There is an even more important beginning than mathematics. This is God.

He endures always and is present everywhere, and by existing always and everywhere he constitutes duration and space. Since each and every particle of space is *always*, and each and every indivisible moment of duration is *everywhere*, certainly the maker and lord of all things will not be *never* or *nowhere* ... God is one and the same God always and everywhere. He is omnipresent not only *virtually* but also *substantially*; for active power cannot subsist without substance. [Newton, I. 1999. *The Principia: Mathematical Principles of Natural Philosophy*, I.B. Cohen and A. Whitman (trans.), Berkeley: University of California Press, 941.]

Forming the foundations of mechanics, Newton refers to the will of God, who builds the universe most rigorously and flawlessly, so that it does not depend on people and their mental abilities. In understanding this project, mathematics is of decisive importance, and not, for example, psychology as a study of human mental capabilities. Without mathematics, one can hardly understand the innermost nature of physics. This line of thought was quite unexpected for the British supporters of the empirical philosophy of John Locke, John Berkeley, and David Hume. The modern researcher understands that Newton attributes to God his understanding of the nature of physical theory. It consists in the fact that, first, several laws must be enough to explain all experimental data. Second, similar causes determine similar effects. Third, the formulations of laws are the results of induction.

As for the laws of mechanics, there are only three of them. Moreover, Newton's first law, according to which "Every body perseveres in its state of being at rest or of moving uniformly straight forward, except insofar as if is compelled to change its state by forces impressed" apparently not a law. It is a definition, clarifying the nature of the inertial reference frame, within the boundaries of which the second and third Newton's laws are fulfilled. The point is that real forces are always the interaction of bodies. Sometimes there is an illusion that there are forces not caused by bodies. For example, passengers in a speeding car feel that a force pushes them against the C-pillar. In reality, there is no such force, for there is no object that would cause it. Thus, one should clearly distinguish between real and imaginary forces. Only real forces fix the inertial frame of reference. Essentially, Newton defines it as a system in which the first law of mechanics is fulfilled. Nevertheless, the other two laws are also fulfilled only in inertial systems. Newton did not escape the logical circle: the inertial frame of reference is determined based on the laws of mechanics, and these laws are by reference to the nature of inertial systems. Newton, explaining the meaning of the inertial frame of reference, had simply to confine himself to explaining the difference between real and imaginary forces.

Newton was adept at the scientific methods of deduction and induction. They allowed him to identify extraordinary scientific positions. He undoubtedly understood the relevance of the methods he used. Unfortunately, in his scientific zeal, he crossed a certain border, claiming that there is the absolute time, space, place, and motion. He strengthened this opinion by considering the experience with a bucket filled with water. If the bucket is set in rotational motion around its axis, then the surface of the water takes on a parabolic shape. He believed that this form makes sense only when introducing the concept of absolute space as the container of all physical objects. This was a critical mistake. Geniuses are wrong too!

Newton was a contemporary of René Descartes, Robert Hooke, Christian Huygens, and Gottfried Leibniz. His theory should be judged against the background of the contribution of these excellent scientists to the development of mechanics. This contribution is significant. Nevertheless, only Newton managed to give mechanics a slender paradigmatic appearance, permeated with conceptual unity.

25. Einstein's breakthrough to the relativistic paradigm in physics

Newtonian mechanics dealt with gravitational and electromagnetic forces, the mechanism of which was not known. It was to be opened. It is instructive how physicists did this. The study of electromagnetic phenomena was more successful than the study of gravity. Putting in mathematical form the results of the excellent experimental work of Michael Faraday by James Maxwell, which was then significantly improved by Oliver Heaviside, led to the creation of classical electrodynamics with its four laws. Despite the mathematical elegance of electrodynamics, it possessed several paradoxical features. The main ones were the following. First, the nature of the substance that supposedly provided the very existence of electromagnetic interactions, including the waves discovered by Hertz, was unclear. Presumably, it was luminiferous ether. Second, the speed of light emitted by a source did not depend on the speed of the source itself. Thirdly, there were doubts about the possibility of the same kind of laws of electromagnetism in different frames of reference.

When dealing with a heap of paradoxes of classical electrodynamics, most physicists tried to explain them by considering the actions of certain forces. This line of thought seemed quite natural, consistent with the maxim of Newtonian physics. Following it, Hendrik Lorentzen suggested that bodies consisting of charged particles, passing through the ether, contract in the direction of motion.

Now a young 25-year-old researcher Albert Einstein enters the arena of great physics. He bases his thinking on two principles, firstly, the principle of relativity, according to which in all inertial frames of reference all the laws of electrodynamics are the same; secondly, the principle of short-range action: the speed of light in a vacuum, being constant, does not depend on the speed of its source. Something very significant is happening. The physical theory takes on a harmonious form, several physical paradoxes disappear, and new phenomena are discovered. You no longer need to worry about absolute space, absolute time, and ether; they do not exist. Revealed the relativity of simultaneity, lengths, and durations, and the huge potential of the internal energy of a body at rest ($E_0 =mc^2$).

Einstein thinks in a fundamentally different way than most of his colleagues. They are primarily concerned with finding the causes of physical phenomena. They consider them self-sufficient. Einstein starts with principles; the causes he is ready to consider only in the context of principles and laws.

Surprisingly, the reasons are not self-sufficient. Many researchers are amazed at the relativity of extent and duration. It seems to them that the relativity of the mechanical speed of movement is customary, but not the relativity of the lengths of bodies and the durations of the processes. It is difficult for many positivist writers to understand that causation depends on the principles and laws of theories. The dependence of measurement methods on the principles of theory is difficult to perceive. It has been repeatedly argued that the relativity of lengths and durations is not a real fact, but just a paraphrase of their methods of measurement. Meanwhile, the method of time measurement used in relativistic mechanics, which is surprising to many, is a direct expression of the short-range principle.

It is often argued in the physics literature that relativistic mechanics is a simple theory. Its understanding, they say, is quite within the power of schoolchildren. On the other hand, now and then venerable physicists cannot agree in understanding some of the conclusions of relativistic mechanics, for example about of relativity of body mass.

Returning to Einstein, I will note that thinking through principles is very characteristic of him. This characteristic of Einstein's thinking does not save him from mistakes. Quite indicative in this respect is his creation of a relativistic theory of gravitation. His main achievement, undoubtedly deserving of the highest estimate, consisted in the expression of the dependence of the space-time tensor on the energy-momentum tensor. Another innovation of Einstein were the concepts of the special and general principle of relativity. According to the special principle of relativity, the laws of physics are the same in all inertial reference frames. According to the general principle of relativity, they are the same not only in all inertial but also in non-inertial reference frames. The fact is that in non-inertial frames of reference one has to distinguish between true forces from untrue ones. Principles, by definition, refer only to true forces. Thus, there is no general principle of relativity as such. Einstein created the relativistic theory of gravitation, not general theory of relativity. He also created relativistic electrodynamics, but not the principle of special relativity.

Thinking through principles should be learned primarily from Einstein. However, scientific vigilance should be observed. Thinking through principles is relevant only when it is organically combined with all stages of conceptual transduction. Einstein fulfilled this rule when applied to relativistic, but not to quantum mechanics. He insisted that probabilistic representations do not fit the nature of quantum mechanics. There was no data for such a statement.

There have also been cases where Einstein over-relied on dubious experimental data. The cosmological model of the structure of our Universe proposed by him was based on the erroneous idea that galaxies form a fixed horizon about the earth. In this regard, Einstein introduced the so-called λ-term into the equations of gravitation. If he had not done this, then most likely he would have become the discoverer of the phenomenon of the scattering of galaxies.

Einstein's incidents show that it is extremely difficult consistently to demonstrate exemplary scientific competence. Failures inevitably happen, and then even geniuses like Einstein find themselves in metaphysical ravines.

26. Bohr's strange philosophizing

There is hardly an intellectual, who has not heard of the debate between Niels Bohr and Albert Einstein over the foundations of quantum mechanics. Einstein, not sure of the solidity of probabilistic concepts, transferred the ideals of classical physics to quantum mechanics. Bohr, objecting to such a transfer, nevertheless believed that quantum mechanics, with all its innovation, should reproduce all the achievements of classical physics. Both Bohr and Einstein demonstrated the same direction of thought, they went from classical physics to quantum mechanics. Another line of thought, namely the movement from quantum mechanics to classical physics, did not attract the attention of the two geniuses of science as much as the first trend. Meanwhile, there are good reasons to believe that the path of thought quantum mechanics → classical mechanics is more relevant than the transition from classical mechanics to quantum mechanics. To prove this statement, let us turn to intratheoretical transduction.

When physicists first encountered quantum phenomena, only classical physics was at their disposal. They could foresee the course of possible phenomena solely based on classical concepts. But experiments supplied facts that could not be explained through classical physics. Thanks to their creativity, physicists have created a new theory. At the same time, they managed to find out why classical physics did not possess the potential of a new theory. Physical proof went from a new theory to an old one, and not from an old, obsolete, to a new theory. The new theory is the key to understanding the old theory; an outdated theory is not the key to understanding a new theory. The new theory is not created in the image and likeness of the old theory. Otherwise, there would be no need at all to create a new theory.

It is time to turn directly to Niels Bohr's position. He actively promotes three fundamental provisions, now and then calling them principles, this is, firstly, the principle of correspondence, secondly, the need to formulate the results of measurements in the language of classical physics, and thirdly, the principle of complementarity.

> The correspondence principle expresses the tendency to utilize in the systematic of the quanta theory every feature of the classical theories in a rational transcription appropriate to the fundamental contrast between the postulates and the classical theories. [Bohr, N. (1925) Atomic theory and mechanics. Nature, Supplement 116 (2927), 849.]

> It is decisive to recognize that, however far the phenomena transcend the scope of classical physical explanation, the account

of all evidence must be expressed in classical terms. The argument is simply that by the word 'experiment' we refer to a situation where we can tell to others what we have done and what we have learned and that, therefore, the account of the experimental arrangement and of the results of the observations must be expressed in unambiguous language with suitable application of the terminology of classical physics. [Bohr, N. (1958) Atomic Physics and Human Knowledge. New York: John Wiley, 39.]

[E]vidence obtained under different experimental conditions cannot be comprehended within a single picture, but must be regarded as complementary in the sense that only the totality of the phenomena exhausts the possible information about the objects. [Ibid, 40.]

In my opinion, all three principles are untenable as the foundations of quantum mechanics. To realize the principle of correspondence, one must know the content of classical physics, which is not outdated. It is established based on the principle of mature knowledge. After that, there is no need for a correspondence principle.

Bohr's conviction that experimental data should be expressed in the language of classical physics, and not quantum physics, also misses the mark. During his lifetime, it was not known that classical physics is consistent insofar as it deals with mixed states, the nature of which is explained by quantum physics. This means that the language of experimental physics is the language of quantum mechanics.

As for the principle of complementarity, it seems to be innovative only when the wave-particle dualism from classical physics is at the forefront. Contrary to it, the principle of complementarity argues that both sides of dualism should be united. This proposal for quantum physics is not consistent. The fact is that there is no dualism in it. What is not does not need to be overcome. All experimental data of quantum mechanics form a unity insofar as they are in full agreement with the foundations of the theory, in particular, with the principle of least action and the representation of the states of physical systems by means of wave functions.

I turned to the work of Niels Bohr not by chance. The fact is that researchers, both philosophers and physicists, often take for the philosophy of physics not what is its actual content. In particular, the status of the philosophy of quantum mechanics is often associated precisely with the content of Bohr's principles discussed above. Nevertheless, they, as I have shown, do not express the true content of quantum mechanics. The philosophy of quantum mechanics begins with an examination of the principles of quantum mechanics itself, with an analysis of the concept of a wave function and an operator representation of physical quantities. Bohr considered this problem in a meaningful way. His analyzes were very informative. Nevertheless, his persistent attempts to impose the authority of classical physics on quantum mechanics do not fall within the realm of the true philosophy of physics.

27. Max Born case

It is well known that understanding a new theory is always fraught with some difficulties. They are the more, the more radically the new theory differs from its predecessor. Quantum mechanics is more complex than classical mechanics, and quantum field theory is more complex than quantum mechanics. At the same time, quantum mechanics differs from classical mechanics more than quantum field theory from quantum mechanics. It is not surprising, therefore, that it was precisely the understanding of quantum mechanics that caused the greatest difficulties for physicists. This was especially true of its probabilistic nature.

Before the advent of quantum mechanics, physicists believed that probabilistic representations were appropriate only insofar as the information at their disposal was not accurate enough. After their refinement, the need for probabilistic representations disappears. This opinion turned out to be erroneous as applied to quantum mechanics. It turned out that probabilistic representations express the nature of quantum mechanical phenomena most organically. Max Born developed this understanding of the nature of physical phenomena.

His undoubted innovation was recognized relatively late. Physicists who were the first to present the equations of quantum mechanics became laureates of the Nobel Prize in physics 5-7 years after their discoveries. Werner Heisenberg in 1932, Erwin Schrödinger and Paul Dirac in 1933. Max Born became the Nobel Prize laureate in physics only in 1954, 26 years after his discovery. The Born case is very instructive in terms of understanding both the nature of physics and science as a whole. It is especially interesting to follow the logic of Born's discovery, which he explained in his Nobel lecture.

In 1900, Max Planck, considering the thermal radiation of a solid, concluded that to avoid getting meaningless results, it should be recognized that the energy of electromagnetic waves is emitted in discrete portions proportional to the frequency of radiation. Frequencies of radiation indicate the wave nature of light, and radiation in portions indicates corpuscular properties. Therefore, we have wave-particle dualism.

In 1905, Albert Einstein, proceeding from Planck's ideas, developed the theory of the photoelectric effect, the emission of electrons by a substance under the action of light waves. In 1913, Niels Bohr explained the structure of the atom, assuming that electrons are in stationary orbits. When electrons jump from one orbit to another, they emit a quantum of energy. The question arose about the probability of radiation. By the content of Einstein's theory of the photoelectric effect, this probability should be proportional to the square of the amplitude of the light wave. In 1923-1924 Louis de Broglie substantiated the wave-particle nature of all particles, not just photons. In 1925, Erwin Schrödinger and Werner Heisenberg present the equations of quantum mechanics. Schrödinger uses the concept of a wave function. Heisenberg abandons the concept of electronic orbits because they are not observed in experiments. His focus is on the statistics of experimental data. It is reasonable to assume that it is connected in a certain way with the wave function. Howbeit the wave-particle dualism with its contradictory character remains in force.

Max Born saw a way out of this difficult situation in treating the wave function as a description of the states of quantum systems. It allows predicting in a probabilistic form the totality of the values of all features of quantum systems. Born's concept of probability does not appear by chance, but as a concept necessary to overcome the wave-particle dualism. The concept of a wave function in quantum mechanics or field operators in quantum field theory allows one to judge physical reality not directly, not 'head-on', but indirectly, predicting experimental data and highlighting invariants in them that are usually associated with particles. Born assessed the lessons of his creativity.

> The matter concerns the borderland between physics and philosophy... The lesson to be learned from what I have told of the origin of quantum mechanics is that probable refinements of mathematical methods will not suffice to produce a satisfactory theory, but that somewhere in our doctrine is hidden a concept, unjustified by experience, which we must eliminate to open up the road. [Born, M. (1954) Nobel Lecture. The statistical interpretation of quantum mechanics, 256, 266. Available at: *https://www.nobelprize.org/prizes/physics/1954/born/lecture/*]
> All great discoveries in experimental physics have been due to the intuition of men who made free use of models, which were for them not products of the imagination, but representatives of real things. How could an experimentalist work and communicate with his collaborators and his contemporaries without using models composed of particles, electrons, nucleons, photons, neutrinos, fields and waves, the concepts of which are condemned as irrelevant and futile? [Born, M. (1953) Physical Reality. The Philosophical Quarterly 3:10, 140.]

I accept Born's conclusions with pleasure. However, his reference to the beneficialness of intuition seems to me superfluous. He continued the chain of outstanding discoveries of Planck, Einstein, Bohr, De Broglie, and Schrödinger. Despite these discoveries, their theories were conflicting. None of them refuted wave-particle dualism. Each of them failed to present the cycle of intratheoretical conceptual transduction without controversy. This meant that they did not properly implement the final method of this cycle, abduction.

It is precisely abduction, which presents the novel cycle of conceptual transduction in a final harmonious form, devoid of contradictions, including taking into account experimental data. Born revealed the mechanism of his creativity. Where there is a mechanism, intuition, with its emphasis on instant insight, is out of work.

I still have to explain the delay in recognizing Born's undoubted merits. In my opinion, his conception was much more comprehensive and substantive than the theories of Heisenberg, Schrödinger, and Dirac. He managed to present the conceptual originality of quantum mechanics in the most distinct form. According to the memoirs of Born's contemporaries, he felt a sense of resentment in connection with the late award of the Nobel Prize to him. Born had more grounds for pride than for resentment. As Sergei Yesenin, a Russian poet once said, "Great things are seen at a distance".

28. The war to understand physics

The development of physics has more than once puzzled scientists. Many of them even thought that they ceased to understand it. After quantum mechanics, quantum field theory became the next contender for the status of a difficult to understand science. Until the late 1940s, physicists tried unsuccessfully to achieve the desired clarity in understanding it. This circumstance manifested itself most clearly in the calculation of the values of the parameters of elementary particles. Instead of finite, infinite quantities appeared that had no physical meaning. Infinite values of physical parameters have never been recorded in experiments. Another trouble was the growing doubts about the legitimacy of using seemingly quite adequate mathematical means, in particular, differential calculus with its concept of infinitesimal quantities. Is it legal to use them? Physicists knew that, according to calculations, the self-energy of particles with a charge, for example, electrons, depends on their size. If the electron is likened to a point, i.e. to consider him infinitely small, then his energy must be recognized as infinitely large. Also, one should take into account the effect on the electron of the field in which it is located. According to calculations, it is also often infinitely large. However, it essentially depends on the constant (coefficient) of interaction, the value of which depends on their type.

Therefore, physicists found themselves in a situation of confrontation with infinities. The most radical way to get rid of infinities is to subtract other infinities from them. In this regard, physicists have developed an ingenious renormalization technique. To neutralize infinities, counter-terms of a similar form are added to the equations, in front of which are some constants (regulators). It turns out that their choice is possible such that all infinities disappear, and the main parameters of the theory will be finite. If the choice of a finite number of counter-terms neutralizes all infinities, then the theory is by definition renormalizable. To the satisfaction of physicists, it turned out that quantum electrodynamics can be released from all infinities by modifying the values of only two parameters, the mass, and charge of the particles. Their values should be determined based on experimental data. One should trust not only the formalism of physical theory and mathematics but also experimental data.

> The war against infinities was ended. There was no reason anymore to fear the higher-order terms. The renormalization took care of all infinities and provides an unambiguous way to calculate with any desired accuracy any phenomenon resulting from the coupling of electrons with the electromagnetic field. It was not a complete victory, because infinite counter-terms had to be introduced to remove the infinities. Furthermore, the procedure of eliminating infinities could be carried out only by renormalizing successively at each step of the perturbation expansion in powers of the coupling parameter. It still is not clear whether this method leads to a convergent series. It is like Hercules's fight against Hydra, the many-headed sea monster, which grows a new head for every one cut off. But Hercules won his fight and so did the physicists. Sidney Drell characterized the situation most aptly as "a peaceful coexistence with the infinities". [Weisskopf, V.F. (1981) The development of field theory in the last 50 years. Physics Today 34(11), 79.]

As you can see, the situation is somewhat ambiguous. Even Richard Feynman, one of the founders of the renormalization procedure, expressed doubts about the legality of it.

> I think that the renormalization theory is simply a way to sweep the difficulties of the divergences of electrodynamics under the rug. I am, of course, not sure of that. [Feynman, R.P. (1966) The development of the space-time view of quantum electrodynamics. Science 153(3737), 707.]

However, there is an obvious fact for optimism. It was possible to preserve the achievements of the quantum field formalism and to understand better the physical side of the issue. It consists of interpreting the coefficients of the counter-terms following the chosen type of regularization. In this regard, a large field for creativity opens up. It opened only after the obstacle in the form of seemingly meaningless infinities was overcome.

The renormalization concept makes it possible to formulate several philosophical conclusions. First, it raised the question of the relationship between physics and mathematics in a new way. Mathematics is not autonomous; any of its positions need physical interpretation. Second, the concept of renormalization revealed the organic unity of deductive and adductive methods. The formulation of the theory without taking into account the experimental data is impossible in principle. Third, it turned out that the existing physical theories refer to the low-energy level of particles. With the transition to a high-energy level, due to a change in the intensity of interactions, the need for new theories arises. Fourth, at low energy of elementary particles, the atomic hypothesis prevails; they can be considered as having no structure. Fifth, the renormalization concept limits the claims of the universal approach. The fact is that one has to take into account the types of interactions and the presence of different energy levels in them. Sixth, physical theories weaken the claims of naive realism. Ontology changes with the development of physical theories. Ontology does not precede theory. Seventh, the unity of physical knowledge is increasing. This is not surprising, because we constantly have to deal with its variability, in connection with which it is necessary to link together the historical stages of the development of physics. Eighth, the unity of physical knowledge makes high demands on the philosophical training of physicists. Its presence makes it easier to carry out various generalizations.

29. A disagreement among physicists over the interpretation of quantum mechanics

On the morning of July 17, 1936, Niels Bohr paid a visit to Erwin Schrödinger, who at the time was a professor at Oxford University. They had a long conversation about the correct interpretation of quantum mechanics. Each of the two Nobel Prize laureates in physics insistently asserted the correctness of his conception. They clearly could not agree with each other. In this regard, a reasonable question arises about the possibility of the successful teaching of quantum mechanics to students. One gets the impression that each of the disputing parties skips past the truth. The piquancy of the situation lies in the fact that outstanding physicists in no way explain the possibility of generalizing different points of view. It was not without difficulty that I found in the articles of one of them a judgment deserving attention regarding the pluralism of opinions. In the final part of his Nobel lecture, R. Feynman made a significant statement

> that a good theoretical physicist today might find it useful to have a wide range of physical viewpoints and mathematical expressions of the same theory (for example, of quantum electrodynamics) available to him. This may be asking too much of one man. Then new students should as a class have this. [Feynman, R.P. (1966) The development of the space-time view of quantum electrodynamics. Science 153(3737), 708.]

Unfortunately, Feynman's assertion explains little. It does not answer several questions. Why exactly should there be different points of view? Should they be combined, and if so, in what way? Is it permissible to assume that different interpretations represent the same theory? Are these interpretations physical conceptions? Is it okay to stick to one single interpretation?

Let us first turn to the status of interpretations of quantum mechanics. It is usually emphasized that they are at the intersection of physics and philosophy. What exactly is at this junction? In my opinion, a metascientific theory. Trying to understand some theory leads to the need to create a theory about it. A theory about a theory is metatheory. Interpretations of quantum mechanics are a family of quantum mechanics metatheories. It is significant that the lack of proper attention to metatheory negatively affects the understanding of the theory. From this point of view, interpretations of quantum mechanics are not only desirable but also necessary.

The total number of interpretations of quantum mechanics is unknown. The most famous is the Copenhagen interpretation of Bohr, Heisenberg, Born, and several other prominent physicists, the many-worlds interpretation of Hugh Everett III, and the ensemble interpretation of Einstein and Schrödinger.

According to the Copenhagen interpretation, the wave function refers to a separate quantum object. The measurement causes the collapse of the wave function; hidden parameters are not possible.

According to world interpretation, the universal wave function represents the entire physical universe. The measurement processes form separate physical worlds. By choosing this or that alternative, people do not destroy other physical worlds, the reality of which manifests in interference patterns.

Within the ensemble interpretation, the wave function refers to statistical ensembles, and not to individual particles. The main function is not the wave function, but the density operator.

Different interpretations of quantum mechanics can be combined, for example, as follows. The Copenhagen interpretation is considered the most standard. The strengths and weaknesses of the Copenhagen interpretation are noted. After that, an alternative theory is formulated. The same is done with other interpretations. The chain of alternative concepts forms metatheory of this or that physicist. Strictly speaking, creative impulses push him beyond the limits of one interpretation. If the researcher does not understand this circumstance, then he declares himself a supporter of one of the conceptions.

The presence of different interpretations of quantum mechanics testifies to the extraordinary nature of the metatheory of quantum mechanics. They cannot be combined into one conception. The originality of the creative positions of physicists is inevitably accompanied by the discord of their theories. We must not forget that science is an intersubjective exercise. It inevitably presupposes a rivalry between different metaconceptions. Of course, this rivalry should not be led to strife. The success of metascientific work is facilitated not only by criticism of the theories of opponents but also by the hermeneutic ability to hear another, assimilating the positive aspects of his theory. Of course, the physicist is faced with the need to present his metatheory in a distinct form. If he does not do this, then, willingly or unwillingly, he contributes to the growth of confusion in scientific affairs. There is often an aspiration to define the only correct interpretation of quantum mechanics. As already noted, it does not exist. We are forced to content ourselves with comparing the strengths and weaknesses of various conceptions of quantum mechanics.

Let me end this essay with a piece of advice given to physics students about the chapter on interpreting quantum mechanics by their witty teacher.

Do not read this chapter, especially if you are about to take quantum mechanics. If you still dare to read it, then the author disclaims all responsibility for your mental health, as well as for the mark received on the exam. If contrary to advice, you are still interested in interpretations of quantum ¨mechanics, avoid discussing what you read with the examiner during the time of the exam in theoretical physics. However, if you have to pass an exam and/ or an abstract in philosophy, this chapter can be useful, especially if you also find time for reading the book by V. I. Lenin "Materialism and Empirio-criticism". [Ivanov, M.G. (2012) How to understand quantum mechanics. Moscow-Izhevsk: Research Center "Regular and Chaotic Dynamics", 259. [In Russian].]

When passing the exam to physicists, be vigilant, they, as a rule, have a vague idea of metascientific knowledge. Contrary to the advice of Mikhail Ivanov, one should be just as vigilant at meetings with philosophers. Among them, people who are versed in quantum mechanics are extremely rare. As for Ivanov's reference to Lenin's book, it is especially amusing. Lenin believed that at the beginning of the 20th century, physicists did not give birth to quantum mechanics, but dialectical materialism. Someone who was completely incompetent in the field of physics could only reach this conclusion.

30. Philosophy of chemistry in the struggle for independence from the 'big brother'

After physics, it is time to turn to the nature of chemistry, the science of the properties and transformations of substances consisting of atoms, molecules, and ions. In considering this issue, one should take into account the close connection between chemistry and quantum mechanics. In 1929, the eminent physicist Paul Dirac mentioned physics and chemistry in the same context.

The underlying physical laws necessary for the mathematical theory of a large part of physics and the whole of chemistry are thus completely known, and the difficulty is only that the exact application of these laws leads to equations much too complicated to be soluble. It therefore becomes desirable that approximate practical methods of applying quantum mechanics should be developed, which can lead to an explanation of the main features of complex atomic systems without too much computation. [Dirac, P.A.M. 1929. Quantum Mechanics of Many-Electron Systems. Proceedings of the Royal Society of London. Series A, Mathematical, Physical and Engineering Sciences 123(792), 714.]

It is reasonable to assume that if two sciences have common laws, then they coincide with each other; therefore, chemistry is no different from physics. This line of reasoning presents professional chemists with an unpleasant choice. Either they must abandon their habitual status as chemists, or they do not pay enough attention to the foundations of chemistry, which coincide with the foundations of physics.

Reductionists believe that chemistry boils down to physics. Emergentists believe that qualitative leaps have taken place in the evolution of matter. When applied to chemistry, this means that a chemical has properties that cannot be explained by the principles and laws of physics. In my opinion, both sides are wrong.

All attempts by chemists to discover such phenomena that contradict the principles and laws of physics ended in vain. However, it does not follow from this that chemistry is physics. The peculiarity of chemistry lies in the fact that the sequence of the conceptual mechanisms it considers, starting with the principles and laws of physics, in particular, quantum mechanics and thermodynamics, extend to atoms, molecules, and ions, as the material content of chemical reactions. The chemist goes beyond physics. He overcomes so many specific difficulties that it is somehow inconvenient to think that he is just repeating the exploits of physicists. If we take into account that the modern classification of branches of science takes into account the peculiarities not only of principles and laws but also of various ways of their specification, then chemistry should be recognized as a special branch of science. Another subtlety is that the principles and laws of physics are used in chemistry only insofar as they have chemical relativity. In chemistry, they have chemical and in geology geological relativity.

The special status of chemistry leaves a certain imprint on chemical theories. They, following physical theories, go beyond them. They are advanced theories. These are, for example, two famous theories of the chemical bond, valence bond, and molecular orbital. Both theories begin with the potential of quantum mechanics.

Philosophers of chemistry give much less attention to the concept of theory than philosophers of physics. In my opinion, at least in part, this circumstance is explained by the insufficient attention of the philosophers of chemistry to the status of advanced theories. The lack of proper attention to the concept of theory leads to a misunderstanding of the relevance of the principle of theoretical representation. The practice-centered approach dominates in the modern philosophy of chemistry. It and similar approaches are asserted, in particular, through references to the metaphysics of Marx (Rein

Vihalemm), to American pragmatism (Joseph Earley), to the philosophy of the late Wittgenstein (Romano Harré). References directly to chemistry are especially significant. Roald Hoffmann, the Nobel Prize laureate in chemistry, who deservedly enjoys high prestige among the philosophers of chemistry, cites them, in particular.

> In chemistry, it is clear that making new molecules is a very, very different enterprise from analyzing what is in nature. And that synthesis creates in its practitioner's different ways of looking at the world, in which theory building is not central. Making things is. [Hoffmann, R. (2007) What might philosophy of science look like if chemists built it? Synthese 155(3), 332.]

Like many other authors, Hoffman takes practice beyond theory. He believes the theory is figuring out what is chemical, not what chemists should synthesize. In my opinion, synthesizing new chemicals is nothing more than a kind of chemical experiment, i.e. they refer to the adductive stage of conceptual transduction. Making things must not be excluded from science, because in this case any possibility of interpreting their content is omitted. The main drawback of all adherents of the practice-centered approach is the insufficient analysis of the conceptual nature of chemistry, which is a collection of theories, and nothing else. The practice-centered approach is asserted as a metaphysical proposition. It is not properly substantiated. In addition, it does not directly relate to chemistry as such, but to its interdisciplinary connections. This point is extremely important when discussing the status of the practice-centered approach.

When chemists once again demonstrate their remarkable ability to synthesize new substances, they, as a rule, take into account the requests of representatives of axiological sciences, for example, medicine, technology, economics, art history. The values of the axiological sciences are absent directly in chemical theories. Chemistry has not an axiological content, but axiological relativity. Developing chemical theories, chemists produce new substances that have chemical properties, for example, reactivity to various other substances, corrosion resistance (to water, moist air, and electrolyte solutions), and electronegativity (of elements). The content of chemical theories testifies to the fact that the axiological relativity of chemical properties is not expressed in them. It comes to the attention of the representatives of axiological sciences. So, one should distinguish between 'chemistry as such' and 'chemistry in its interdisciplinary relationship with the axiological sciences'. Chemists deal with the chemical properties and chemical relativity of non-chemical, for example, of physical and technical properties.

31. Dualism: mobilism vs fixism

As you know, the history of our Universe began with some physical particles. Chemical processes became a reality only after the synthesis of chemical elements in the interiors of stars. It took more than 9 billion years before the formation of the Earth and, accordingly, the Earth sciences began. Improving the earth sciences is an exciting chapter in the development of natural science. The Earth is a complex dynamic object consisting of more than a dozen geospheres, in particular, the lithosphere, hydrosphere, pedosphere, atmosphere, biosphere, ionosphere, and magnetosphere. The solid part of the Earth is the object of geology, which consists of hundreds of theories. Of course, researchers are trying to work out some unified theory about them. In this regard, since the 1960s, high hopes have been pinned on plate tectonics. According to this theory, the outer shell of the Earth consists of lithospheric plates that lie on the Earth's mantle and move along it. Plate tectonics explains the experimentally recorded continental drift. The growth history of Plate tectonics' authority is extremely revealing and instructive.

The German geologist Alfred Wegener founded plate tectonics. Trying to explain the similarity of the outlines of the west coast of Africa and the east coast of South America and the presence of paleontological data on the similarity of their flora and fauna, Wegener concluded that the Earth was once a supercontinent, which disintegrated into parts, later separated from each other. There was a weak point in his conception; he could not explain the origins of the forces capable of moving continents. Numerous critics of Wegener's theory have rejected it as insufficiently scientific. Plate tectonics revived in the 1960s after the discovery of ridge networks located in the central parts of all oceans and numerous magnetic anomalies. The movement of lithospheric plates could explain both. Since the 1970s, Plate tectonics has established itself as the main geological theory. Nevertheless, it has an alternative in the form of the theory of geosynclines, according to which the vibrational movements of the crust of the earth lead to the appearance of folded regions, geosynclines. Summarizing the content of the theories of tectonic plates and the concepts of geosynclines, respectively, scientists speak of two geological directions, mobilism, and fixism. Mobility of lithospheric plates and continents as integral formations are recognized only within the framework of mobilism.

The rivalry between the mobilists and the fixists stretched out for a whole century. This fact is surprising, because, according to many researchers, it testifies to a certain misunderstanding of the nature of the dynamics of scientific knowledge. On this score, the Russian geologist Alfred Naimark and the Argentine geologist Pablo Pelegrini develop interesting views.

[No (fixist, mobilist, other) concept, being a model of reality, can, by definition, neither fully correspond to reality, nor completely contradict it. No empirical fact, no experience as a whole, can unequivocally and definitively neither confirm nor refute any concept. A rigorous assessment of the degree of factual compliance is impracticable, and therefore cannot be a practically applicable criterion of the truth, usefulness, and preference of a concept. Only the greater efficiency of the application (not at all unambiguously associated with "greater correspondence" to the facts) gives the new concept wide recognition and a leading role.

[Naimark, AA (2006) Half a century of discussion of fixists and neomobilists: analysis of reality or hypotheses, search for truth or "convenient" theory? Bulletin of Krounz. Earth sciences. 2006. Vol. 8. No. 2. P. 177. (In Russian).]

Naimark, being one of the most prominent critics of the theory of plate tectonics, beliefs that geologist have often achieved success, guided by the theory of geosynclines, even if the facts seemed to contradict it. In my opinion, Naimark is mistaken in many ways. Like many other authors, Naimark separates theory from facts. In this case, it is quite legitimate to argue that there is often no correspondence between theory and facts. In reality, facts are concepts of the theory itself, so they always correspond to it. This statement seems to contradict the practice of scientific research. Meanwhile, it is quite consistent. Let us assume that we have at our disposal a theory T1 containing facts f1i. We are interested in the facts f2k. When scientists say that the theory does not correspond to the facts, they mean that the facts f2k do not correspond to the theory T1. They really may not correspond to it, because they are not part of it. If scientists do the right thing, namely, compare the facts f2k with the theory T2, and then there is always a complete agreement between theory and facts. Contrary to Naimark, theories contradicting the facts are rejected. However, the conformity of theory and its facts is not a criterion for the success of a theory. It is clear why the indicated correspondence cannot be greater or less, it is and nothing more. The criterion for the success of a theory is its ability to initiate the development of league-theories. It is in this respect that plate tectonics have won their competition with geosynclinals theories.

Let me introduce the floor to Pablo Pellegrini.

> [I] analyse continental drift debates from a different perspective which is based on *styles of thought*. I'll argue that continental drift debate took much longer than it was usually recognized with two styles of thought coexisting for hundreds of years. These were fixism and mobilism and they were always confronting their evidence and interpretations and functioning as general frameworks for the acceptability of a specific theory. Therefore, this text aims to bring much broader sociological elements than usually involved in the analysis of the continental drift theory. [Pellegrini, P.A. (2019) Styles of Thought on the Continental Drift Debate. Journal for General Philosophy of Science 50(1), 85.]

Pellegrini believes that the rivalry of theories always takes place under the auspices of certain styles of thinking. This circumstance complicates the process of rivalry between theories. It was not the case that the theory of plate tectonics itself gave rise to mobilism as a geological style of thinking. The development of the theory initially took place under the influence of two styles of thinking, one of which emphasized the variability of phenomena, and the second – on their constancy. When applied to geology, this means that advanced scientists have argued not only the theory of plate tectonics but also mobilism in its opposition to fixism. Fixist attitudes held back the acceptance of plate tectonics.

In my opinion, Pellegrini is right in many ways. The process of developing a theory is never easy. It is not limited to just a generalization of facts but involves carrying out a full-fledged intratheoretical and intertheoretical conceptual transduction, taking into account the ratio of independent and auxiliary theories, including the achievements of the philosophy of science. Therefore, nothing is surprising in the fact that the formation and approval of the theory stretch over many decades, especially when members of the scientific community feel insecure in the field of philosophy of science. Unfortunately, this undesirable situation is in all sciences, including geology. As for the concepts of mobilism and fixism, they are an attempt to express the content of the philosophy of geology. This attempt is rather clumsy because the specified content is not presented systematically.

It is striking that the pair of mobilism vs fixism is shaped in the form of dualism. In modern studies of metamorphic complexes or interactions of geological shells, geologists have quite convincingly shown that mobilism and fixism refer to certain limiting states.

32. Quest for the nature of biological theory

Biology closes several natural branches of sciences: physics – chemistry – Earth sciences – biology. Of course, it occupies a special place in this series. Biology differs from its neighbours largely than they are among themselves. Physics, chemistry, and Earth sciences deductively begin with the principle of least action. In each of these sciences, the principle of least action expresses precisely its specificity. In physics, it correlates with the interactions represented by quantum field theory. In chemistry, the principle of least action expresses the certainty of chemical reactions. In the Earth sciences, the expansion of the principle of least action continues, which this time expresses the features of the fluxes of chemical-density differentiation of matter occurring in various geological shells. Deductively biology begins not with the principle of least action, but with the principles of molecular genetics.

As for the principles of biology, there is no complete clarity on this score. In my opinion, three principles are of decisive importance, namely, those of Crick, Johannsen, and Darwin. The Crick principle sets the direction of biological deduction in the foundations of theories: DNA → RNA → proteins. Johannsen's principle, according to which the genotype determines the phenotype, expresses in an integral form the entire completeness of biological processes. Finally, one cannot fail to mention Darwin's principle of natural selection, which expresses the correlation of organisms with the external environment. In my opinion, the scientific achievements of Crick, Johannsen, and Darwin are so significant that it is quite reasonable to call them by their names the principles of biological theories.

Crick's principle leads many researchers to reductionism. Francis Crick himself argued, "The ultimate aim of the modern movement in biology is to explain all biology in terms of physics and chemistry". (Crick, F. (1966) Of Molecules and Men. Seattle: University of Washington Press, 10.) At the heart of this claim is the belief that DNA and RNA belong entirely to the realm of physics and chemistry. At the same time, reductionists do not take into account that, as applied to biological mechanisms, they signify their deductive beginning. All stages of the process form a single whole. Different principles, not the same, lead to different consequences. Not physical or chemical, but exclusively biological principles cause all biological, in particular, phenotypic consequences. In biology, DNA is a biological concept.

In studying biological processes, researchers have to deal with mechanisms that range from DNA to numerous, including mental, linguistic, and phenotypic characteristics of organisms and their active actions. These mechanisms are extremely multi-link, which naturally makes the process of cognition extremely difficult. In terms of the study of multilink processes, biology is perhaps the champion among all modern sciences. Despite this circumstance, biology has the same conceptual and methodological structure as other sciences. What I mean is that the theory of conceptual transduction is as relevant to biology as it is to other sciences. Conceptually, principles, laws, and variables guide biology, like other sciences. Methodologically, it is not able to do without the methods of intratheoretical and intertheoretical transduction. I believe this remark is relevant insofar as many researchers tend to see biology only as of exceptional conceptual and methodological phenomenon.

In choosing the topic for this essay, I tried to find the most painful point in the philosophy of biology. After special research, I concluded that this is the understanding of the nature of biological theory. On this score, Root Gorelick provides interesting evidence.

> Theory is vital in science, including evolution and ecology, yet is seldom defined. Theory is often amorphously described as anything abstract or mathematical, or is simply defined as the opposite of empirical work. By contrast, I explicitly define theory as the formation of testable hypotheses, while defining empirical work as hypothesis testing. …
>
> Most practitioners of theory never ask what theory is. They simply have some gestalt for what does and does not constitute theory. Most journals containing the words 'theory' or 'theoretical' in their titles do not define theory in their 'aims and scope' nor in their 'instructions to authors'.
>
> Maybe I am naïve, but I disagree with Suppes … when he said, "If someone asks, 'What is a scientific theory' it seems to me that there is no simple response to be given." Hypothesis formation, where a hypothesis is a simple declarative statement, seems to be an unambiguous definition of 'theory'. [Gorelick, R. (2011) What is theory? Ideas in Ecology and Evolution 4(4), 1, 7.]

In my opinion, in her critical attitude to the existing definitions and understandings of the nature of the theory, Gorelick is certainly right. However, her proposal to consider the theory as a declarative statement, which must be tested seems to me clearly insufficient. The main drawback of her position is that she does not have a concept within which the nature of the theory would find its organic expression. She cites numerous authorities, but ultimately it does not come to form a concept worthy of being an alternative to the theory of conceptual transduction. According to this conception, the theory is the management of the concepts of entities, principles, laws, and variables through the methods of deduction, adduction, induction, and abduction. Defining a theory as just a hypothetical statement greatly impoverishes it.

It is widely believed that a theory is formed as a hypothesis, and then, because of its successful testing, turns into a valid theory that allows predictions to be made is an error. The cycle of intratheoretical transduction always begins with a hypothetical theory and just as adamantly ends with a valid theory. Nevertheless, if a new cycle of cognition takes place, then the theory appears again as a hypothetical conception. The cognition cycle is the transformation of a hypothetical theory into a valid conception. A hypothetical theory matches the nature of science no less than a valid theory. An attempt to dispense with the dynamics of hypothetical and valid theories or to absolutize the meaning of one of them is flawed. The nature of science, including biology, is not in testing hypothetical theories, but in ensuring the full-fledged dynamics of hypothetical and valid theories.

Werner Callebaut, noting the difficulties in comprehending the nature of biological theory, decided that they could largely be circumvented by placing the focus not on 'theory' but 'theorizing'.

> Although 'theory' has been the prevalent unit of analysis in the meta-study of science throughout most of the twentieth century, the concept remains elusive. I further explore the leitmotiv of several authors that we should deal with theorizing (rather than theory) in biology as a cognitive activity that is to be investigated naturalistically. [Callebaut, W. (2013) Naturalizing theorizing: beyond a theory of biological theories. Biology Theory 7(4), 413.]

The program offered by Callebaut is the full use of mathematical modelling, the transition from explanation to understanding, the use of analogies and metaphors, as well as paraconsistent logic and ecological rationality. The task is to develop theoretical biology for the future.

In my opinion, replacing the term 'theory' with the term 'theorizing' is not a productive step. According to the linguistic norm, the noun and the verb always accompany each other. There is no theorizing without theory and theory without theorizing.

Callebaut emphasizes the nature of theorizing, which must be naturalistic. It means that it should be scientific. If so, then it is not without surprise that I state that the extensive article does not mention scientific methods, in particular, deduction and induction, as well as adduction and abduction. Without them, the very understanding of theorizing turns out to be extremely impoverished.

Callebaut's desire to be "beyond a theory of biological theories" seems to be extremely strange. In biology there is nothing but theories, therefore, in principle, it is impossible to be outside of them.

Like Callebaut, many authors seek to dispense with the concept of 'theory'. A frequently used trick is to speak of mathematical models instead of theory. Mathematical models of biological phenomena express some, but not all, features of biological theories. As a rule, consideration of the relationship between biology and mathematics leaves much to be desired.

I see the only reason for the protracted search for the nature of biological theory in the insufficient development of the philosophy of biology. This statement is in no way a belittling of the remarkable achievements of biologists, in particular, in terms of the development of evolutionary theory, genetics, and, especially, modern synthesis. These successes do not eliminate the need to develop a philosophy of biology.

33. Is technology a science?

Modern science consists of more than two dozen branches of science. The authors who write about them are attaching branches of technology to them, which is a very strange thing. Placing technology just next to branches of science, they do not recognize the scientific status, for example, electrical and radio engineering. On what grounds? Walter Vincenti explains the situation as follows.

> Engineers use knowledge primarily to design, produce, and operate artifacts. …Scientists, by contrast, use knowledge primarily to generate more knowledge. (W.G. [Vincenti, W.G. (1990) What Engineers Know and How They Know it: Analytical Studies from Aeronautical History. Baltimore (MD): Johns Hopkins University Press, 226.]

According to Vincenti, engineering, initially acting as applied science, nevertheless, does not limit itself to scientific methods, but adds engineering methods to it that are different from scientific methods. In my opinion, this argument misses the mark. There are no methods in engineering that go beyond scientific methods. This applies to the design of the artefact, and its manufacture, and its use. The design method is a deduction. The method of making an artefact is adduction. The engineer, after carefully selecting the parts that are necessary to produce the artefact, must finally produce it. A scientist acts similarly at the stage of preparing an experiment. Artefact production is the fact-making stage. Using the induction method, the engineer determines to what extent he has succeeded in realizing his idea. According to the inductive result, he correlates his original design principles. What I have described is nothing more than a cycle of intertheoretical conceptual transduction.

Let us now turn to the actions of the user of an artefact, for example, a car. Once again, using the car as a means of transportation, the user assesses the degree of its suitability. By it, he hopes to make some route one way or another. This is the stage of prediction carried out using deduction. Riding itself is the fact-making stage. Comprehension of the facts is carried out by the induction method. Finally, the user, summing up the trip, is guided by new information about his car, if necessary, giving it to repair.

Scientist summarizes the information he receives from engineers and users. By them, he issues new recommendations. Methodologically, they refer again to the stages of deduction, adduction, induction, and abduction. Both the scientist, in this case, the technologist, and the engineer, and the user use theories, the content of which includes artefacts. They indeed have different goals. Nevertheless, it is extremely important that in conceptual and methodological terms, their theories coincide. This coincidence forces us to consider technology, including engineering, a branch of science, and nothing more. There is absolutely no reason to distinguish branches of technology from branches of science.

As for the originality of technology as a branch of science, it finds its expression in the corresponding principles, in particular, in the principles of productivity, safety, efficiency, reliability, and maintainability. There are no such principles in any other branch of science, in particular, in physics or economics.

Technology is one of the axiological branches of science. The point is that all the concepts and, accordingly, the methods of these sciences are the preferences of people, i.e. values. The physical principle of least action is not a human value. He does not depend on people. The performance-enhancing principle of technical artefacts is a human preference that they invented and that they can abandon if necessary.

Nature is such that it can represent the values of people, technological, medical, economic, art history, and others. People make full use of this wonderful feature. As you can see, the nature of values is very peculiar. They do not have impulse-energy, or, as is often expressed, bodily characteristics. Values cannot be photographed. It seems that values do not exist, being just ghosts. Given their importance in human life, this view is wrong. Values exist as representations of natural objects and processes. Natural objects by themselves do not represent any value. They acquire this trait because people purposefully manipulate them.

According to modern views, technology consists of about 150 disciplines and is the most disciplinary-rich branch of science (compare only about 60 disciplines in medicine). This fact indicates that people are extremely energetic in using the opportunity to present technological values by natural materials. They do it for their development. Modern people cannot imagine themselves without cars, aeroplanes, household appliances, and computers. Against this background, the position of the German philosopher Martin Heidegger looks very curious, who believed that technology limits a person.

> Everywhere we remain unfree and chained to technology, whether we passionately affirm or deny it. But we are delivered over to it in the worst possible way when we regard it as something neutral; for this conception of it, to which today we particularly like to do homage, makes us utterly blind to the essence of technology. [Heidegger, M. (1977). The Question Concerning Technology and Other Essays. Translated by William Lovitt, New York: Harper & Row, 311.]

Martin Heidegger extolled the creative nature of man, believing that technical creativity is not part of it. The Russian philosopher Peter Engelmeier, on the contrary, believed that it is in technology that a person realizes his highest creative abilities. In my opinion, technology is one of the branches of science. It is wrong to both deny its fruitful value for people and extol it. Of course, the development of technology must be under constant ethical control.

To conclude this essay, let me criticize those scientists who consider technical artefacts as entities of technology. People, not artefacts, are the essence of all axiological theories. Economists are the essence of economics, not goods and services. Technologists are the essence of technology, not technical artefacts. They invent and use technological values. They impute them to technical artefacts.

34. Agricultural sciences towards a philosophy of science

Agricultural sciences are a complex of sciences that provide the most efficient way for the production of food, feed, fibre, fuel, and other goods by the systematic raising of plants and animals. Like technology, agricultural sciences coexist with natural science. Nevertheless, imposing on it the stamp of its originality, agricultural sciences becomes an independent axiological branch of science. Agricultural sciences act as a representation of natural science, primarily of biology. This happens at the initiative of people. Biology does not contain the potential that would cause the spontaneous emergence of agricultural sciences, in particular, plant growing and animal husbandry. People are loading biology with valuable content by the principles of maximizing the yield of grain crops and the productivity of various animal species. There are no such principles in biology. They begin to act thanks to the purposeful activities of people. This is why the frequent attempts to reduce the agricultural sciences to biology and chemistry invariably fail.

Many authors, trying to avoid erroneous judgments, distinguish between agriculture and agriculture sciences. The sciences deal with agriculture as a whole, which in terms of volume significantly exceeds the agriculture sciences. In my opinion, this idea needs to be adjusted. Based on the principle of theoretical representation, agriculture is nothing more than an object representation of a theory. Agriculture is a theoretical phenomenon and should not be contrasted with agriculture sciences.

Already ancient philosophers, in particular, Aristotle, Epicurus, and Lucretius realized the need to comprehend the unity of agriculture sciences and philosophy. However, to this day, the ways of realizing this unity are largely questionable. Lindsay Falvey, perhaps like no one else, seeks to consider the relationship between philosophy and agriculture sciences as comprehensively as possible.

> [M]odern philosophy can be based on science's methods by generalizing from the outputs of science to propose a wider theory, and it can also explore other methodologies of science and apply them where possible. Within this wide view, the principle branches of philosophy become those of scientific philosophy and include; aesthetics, epistemology, ethics, logic, metaphysics, and various specialist focus areas. Other lists exist, but this is sufficient for the current discussion [...]. [Falvey, L. (2020) Agriculture & Philosophy. Agriculture sciences in Philosophy. Songkhla, Thailand: Thaksin University Press, 34.]

My approach is fundamentally different from Falvey's. The subject of my interest is the philosophy of agriculture sciences as such.

Aesthetics, ethics, logic, and metaphysics are of little use in designing it. Logic and ethics are not philosophical sciences. Metaphysics is unscientific by definition. Aesthetics belongs to art history; it is also far from the agriculture sciences. As for epistemology, it is an organic part of the philosophy of science, in particular, the philosophy of agriculture sciences. Thus, in my understanding, the philosophy of agriculture sciences depends directly on the content of these sciences, expressing their conceptual and methodological nature.

At this point, unfortunately, I must note that authors speaking on behalf of philosophy, as a rule, focus not on the philosophy of the agriculture sciences, but agriculture ethics, which they, moreover, often identify with environmental ethics. The mentioned ethical systems, of course, are actual; however, they should not overshadow the philosophy of agriculture sciences.

There is no room in this essay for a detailed presentation of the philosophy of the agriculture sciences. I will only note that, according to my research, the theory of conceptual transduction is no less relevant for it than for any other special philosophy of science, for example, the philosophy of technology. About the agriculture sciences, all methods of intratheoretical and intertheoretical transduction remain valid. Unfortunately, the central importance of the concept of the theory was not clearly expressed here.

The philosophy of agriculture sciences does not exhaust the entire nature of these sciences for it expresses only their conceptual and methodological structure. For a more comprehensive understanding of the nature of agriculture sciences, it is necessary to take into account the achievements of the sciences auxiliary to it, i.e. not only philosophy of science, but also linguistics, logic, mathematics, computer science, psychology, sociology, pedagogy, political science, jurisprudence, and ethics. Each of these sciences develops aspects of the theory that are inherent in all axiological sciences, including agriculture sciences. For example, agricultural law examines what is permissible and what is not permissible in agriculture. It is, as Neil Hamilton rightly points out, to ensure the continued success of agriculture sciences. He considers it a decisive factor and distinguishes six philosophical questions shaping the future of agricultural law:

1. What is agriculture?
2. Is there a right to farm?
3. Is there a duty of stewardship associated with ownership or use of farmland?
4. What are the limits of private property?
5. Do farm animals have rights?
6. Should plant genetic resources be subject to legal ownership?

The legal issues which flow from these questions, in the form of legislative and administrative rules, legal documents, court disputes, and institutional relations will in many ways define the content of agricultural law in the future. [Hamilton, N.D. (1993) Feeding Our Future: Six Philosophical Issues Shaping Agricultural Law. Nebraska Law Review 72(1), 212.]

In my opinion, these questions are not philosophical. All of them, except the first, are purely legal.

After considering the relationship of agricultural sciences with auxiliary sciences, it is also necessary to consider their relationship with donor sciences, primarily those of them that adjoin them. These are earth sciences and medicine. These links raise the question of the relevance of environmental ethics.

Agricultural ethics differ from environmental ethics. Agricultural ethics is an organic part of agricultural sciences. It is to maximize the welfare of all subjects of these sciences, farmers, scientists, and the population. Environmental ethics is primarily the medical and agricultural ethical relativity of the natural sciences. By themselves, these sciences do not have ethical content. However, without their participation, it is impossible to ensure the successful development of both agricultural and medical ethics. The challenge is to ensure the best possible mix of agricultural and medical ethics.

Unfortunately, many issues of environmental ethics are presented in an unclear form. In this regard, the position of Kristin Shrader-Frechette is of interest.

Given all the problems with ethicists' misguided appeals to ecological laws, to biocentrism, and equal interests, what are possible solutions? At least three come to mind: 1. ethical default rules to use in situations of ecological uncertainty, 2. scientific-case-study environmental ethics, and 3. recognizing human rights against life-threatening pollution. [Shrader-Frechette, K. Ecology and Environmental Ethics. Values and Ethics for the 21st Century. BBVA, 321.]

My position is that the construction of ethical systems should be the result of a careful study of the ethical content of axiological sciences, in our case agricultural sciences. Everywhere, including in agricultural sciences, the understanding of the nature of the relevant theories leaves much to be desired.

35. Search for the nature of medicine

Medicine is an axiological science that consists of a huge variety of theories. The subjects of medicine are not only sick people but also those who are not sick, in the case of prevention of their medical condition. Medical conditions are defined by a system of variables. If these conditions correspond to the norm established by the medical community, then the person is healthy, otherwise sick. Relationships between variables are laws. The basic principle of medicine is the steady improvement of the medical condition of patients.

My research has shown that the conceptual and methodological nature of medicine is revealed most effectively in the theory of intratheoretical and intertheoretical conceptual transduction. All its methods, i.e. deduction, adduction, induction, abduction, as well as the methods of intertheoretical transduction, are characteristic of medicine to a degree no less than, for example, physics and technology. Unfortunately, the medical community does not have sufficient knowledge of conceptual transduction theory. In this regard, even very competent physicians often form views that are insufficiently articulated scientifically. Let me turn to the statement of Alex Broadbent in this regard.

[T]he core medical competence is neither to cure, nor to prevent, disease, but to understand and to predict it. [Broadbent, A.

(2018) Prediction, understanding, and medicine. Journal of Medicine and Philosophy, 43(3), 289.]

Let me evaluate Broadbent's statement from the standpoint of intratheoretical conceptual transduction, the cycle of which includes four stages and, accordingly, four methods. The stages of the cycle replace each other; there is no reason to consider some of the more relevant than others. This means that the concept of 'core medical competence' is out of the question. It is legitimate to argue that the cycle contains four competencies, which together determine the success of the principle of steady improvement in the medical condition of patients. Prediction is one of the competencies realized through the deduction method. Conceptual transduction methods do not form a hierarchy. If desired, you can understand them as a unity of four stages, but not as one of the competencies. As for curing and preventing the disease, it is obvious that one cannot do without them when implementing the principle of medicine. When defining the nature of medicine, first, one should point out its main principle. Otherwise, he cannot avoid metaphysical wanderings with their poorly clarified conceptual content.

Many physicians are not sure about the scientific status of medicine. Now and then an argument is made about the opposition of medicine as a theory to medical practice. In light of the theory of conceptual transduction, the inconsistency of both positions is obvious. Conceptual transduction theory presents the criteria of a scientific theory in the clearest possible way. Every scientific medical theory has concepts of principles, laws, and variables, as well as methods of conceptual transduction. That is why scientists must recognize them as scientific without any reservations. As for medical practice, it can only be carried out when combined with the stages of conceptual transduction. Therefore, contrasting practice and conceptual transduction are untenable. In the theory of conceptual transduction, the nature of scientific theory has been elucidated with the greatest degree of clarity. It is the management of medical concepts of principles, laws, and variables representing patients through intratheoretical conceptual transduction methods.

In modern medicine, there is an increase in attention to the ways of justifying decisions. Researchers dissatisfied with the randomized controlled trial are seeking to develop a more versatile approach, evidence-based medicine. At the same time, they strive to avoid situations of uncertainty, use the best evidence, conduct a systematic study of best practices, and their implementation in practice. Difficulties arise when it is necessary to implement this program, saturated with good wishes. At the same time, there is an urgent need for a systematic theoretical approach, in the absence of which it is hardly possible to achieve harmony of the concepts and methods used.

According to my analysis, the most pressing ambitions of the proponents of evidence-based medicine are in excellent agreement with the theory of conceptual transduction. In its context, proof acts as a multi-tier process that includes experiments, statistics, and much more. The construction of theories acts as a search and finding of the most advanced knowledge. Thus, there is, perhaps, no better basis for the development of evidence-based medicine than the theory of conceptual transduction. In this theory, the question of the nature of principles occupies a significant place. It deserves special consideration.

> Before the early 1970s, there was no firm ground in which a commitment to general principles or ethical theory could take root in biomedical ethics. … [T]he perspective on medical ethics in Europe, America, and most of the world was large that clinicians are obligated to maximize medical benefits and minimize risks of harm and disease for their patients. … No bridge had been built to connect work in philosophical ethical theory with the problems of biomedical ethics, and professional ethics was not conceptualized in biomedicine as an interdisciplinary field. [Beauchamp, T. L. and DeGrazia, D. (2004) Principles and Principlism. In Khushf G. (Ed.). Handbook of Bioethics. Taking Stock of the Field from a Philosophical Perspective. New York, Boston, Dordrecht, London, Moscow: Kluwer Academic Publishers, 55.]

Dissatisfied with the ethical content of traditional medicine, Tom Beauchamp and James Childress considered it necessary to introduce four principles into the composition of medical theories, namely, (1) respect for autonomy, (2) nonmaleficence, (3) beneficence, (4) justice.

In my opinion, the discussed authors misunderstand the relationship between medicine and ethics. They interpret it as an interdisciplinary relationship and believe that it is quite legitimate to supplement medicine with concepts of ethics. Such addition is contraindicated insofar as the concepts of different sciences do not coincide in nature. It is impossible to supplement physical concepts with economic ones, just like economic concepts with physical ones. However, perhaps we have not sufficiently taken into account the nature of ethics. In my opinion, this is the case. There is no need to introduce ethics into medicine; it should be in medicine initially. The ethical project is that in any particular situation, one should seek to improve the condition of all its participants, stakeholders.

Therefore, in a deductive sense, medicine begins with the principle of maximizing the prosperity of all stakeholders, in particular, patients, doctors, and patients' relatives. It is immediately apparent that this principle presupposes all four principles mentioned above as belonging to ethics. It is impossible to imagine its content without respect for autonomy, nonmaleficence, beneficence, justice. This circumstance appears clearly in the joint consideration of the principle of maximizing the prosperity of all stakeholders and the principle of steady improvement in the medical condition of patients.

The second principle excludes harm; improving the patient's condition is incompatible with harming him. The same principle includes beneficence, for improving the patient's condition is beneficence. The principle of maximizing the prosperity of all stakeholders clearly expresses the need for justice. As far as respect for autonomy is concerned, at first glance, it does not contain the principle of a steady improvement in the medical condition of patients. It seems that the improvement of this condition can be achieved without taking into account the position of the patient himself. However, this provision contradicts the principle of maximizing the prosperity of all stakeholders.

Thus, there is no need to introduce into medicine concepts alien to it. Medicine is conceptually self-sufficient. This becomes clear only through a careful analysis of its nature. The widespread lack of attention to the philosophy of medicine indicates the rarity of such an analysis.

36. Endless sociology

Entities of sociology are groups of people, the number of members of which is not limited. It can be a nation, a civilization, or three philatelist friends. Comprehension of the nature of a social group of people, as a rule, begins with the question of the possible construction by individuals with different views a holistic formation, called a group. In my opinion, it is not difficult to answer this question.

Interacting with each other, individuals compare their theories in turn. Of course, a situation is possible when the interlocutors completely deny each other's theories. This is a situation of strife; the French philosopher Jean-François Lyotard has repeatedly discussed it. Much more often, the communication of people leads to a change in their theories. Typically, in a competing theory, interlocutors find fragments to insert into their theory. These are fragments of consent. Intergovernmental negotiation protocols usually contain numerous examples of points of agreement.

Fragments of agreement form group theories. By definition, their subjects are social groups of people, including each of its members. However, is it possible that the same person possessed two theories, individual and group? It is certainly possible. Any person owns numerous theories. Therefore, it is not surprising that he owns both individual and group theory. Every person is a bilateral being. As such, he constantly pits two theories against each other, each time changing their assessment. Those sociologists who give preference to individual theories are representatives of methodological individualism. The German sociologist Max Weber is most often declared the founder of this sociological trend. Those sociologists who give preference to group theories form a camp of supporters of methodological holism, whose leader is the French sociologist Emile Durkheim. It is not hard to see that methodological individualism and holism are two extremes. If a researcher competently uses the theory of intertheoretical transduction, then he can easily overcome the mentioned dualism.

The formation of group theory requires a very careful work of sociologists. They must include concepts in cycles of conceptual transduction. This means that principles, laws, and variables must be highlighted with the necessary degree of clarity. This is far from easy to do. Usually, when defining his position on the principles of the theory, the researcher reckons with many values. Determining which of them are principles and the relationship of subordination between them is a very difficult task. In this sense, the content of the constitutions of developed countries of the world is indicative. The citizens of the Federal Republic of Germany prioritize the values of responsibility and the formation of a united Europe, the citizens of the United States strive, and first, to achieve the improvement of the nation and the building of the rule of law, the citizens of Russia put the values of democracy and federalism especially highly. The three constitutions are three different theories, but they also have similarities, such as a commitment to democracy.

Reflecting on the nature of social theories, Herbert Simon contrasted the criterion of maximizing values with the criterion of their satisfaction.

> *Problem solving by recognition, by heuristic search, and by pattern recognition and extrapolation are examples of rational adaptation to complex task environments that take appropriate account of computational limitations – of bounded rationality. They are not optimizing techniques, but methods for arriving at satisfactory solutions with modest amounts of computation. They do not exhaust, but they typify, what we have been learning about human cognition, and they go a long way toward explaining how an organism with rather modest computational capabilities can adapt to a world that is very complex indeed.* [Simon, H.A. (1990). Invariants of human behavior. Annual Reviews of Psychology 41(1), 11. Italic of Simon.]

In my opinion, Simon is right in many ways. However, he does not take into account two important circumstances. First, by their very nature, in the cycles of conceptual transduction people always strive for their improvement. Secondly, to express this improvement in mathematical form, the operation of maximization or similar is very suitable.

When trying to understand a theory, it is extremely important to understand its mechanisms of functioning. Conceptually and methodologically, they coincide whith the stages and methods of the theory of conceptual transduction. At the same time, it is desirable to detail each of the stages of conceptual transduction in the most concrete way. On this score, supporters of the theory of social exchange and analytical philosophy expressed valuable ideas.

Social behavior is an exchange of goods, material goods but also non-material ones, such as the symbols of approval or prestige. Persons that give much to others try to get much from them, and persons that get much from others are under pressure to give much to them. This process of influence tends to work out at equilibrium to a balance in the exchanges. [Homans, G.C. (1958) Social Behavior as Exchange. American journal of sociology 63(6) 606.]

The selection mechanism in its most basic form is a type of mechanism which links the property of an entity to its reproductive success in a specific environment. The core scheme for the selection mechanism may be summarized as follows:

1. A set of entities, e1 to eN, vary concerning a property, pi.

2. In a certain environment the benefit, bi, bestowed on entity i is a function of its property pi.

3. The value of bi influences the relative frequency of entities with different properties at subsequent periods. (Properties that perform better in the environment will become more frequent.) [Hedström, P., Bearman, P. (2009) What Is Analytical Sociology All About? An Introductory Essay. The Oxford Handbook of Analytical Sociology. Ed. by P. Hedström and P. Bearman. Oxford: Oxford University Press, 7.]

The social exchange that takes place between members of a social group can be quantified if desired. For this, firstly, it will be necessary repeatedly to compare the advantages and disadvantages of competing theories. Secondly, it is necessary to consider the corresponding representations of theories, their embodiment, in particular, in things and their signs.

One of the features of sociology is that the abundance of its disciplines is not represented in any clear form. There are so many of the disciplines that it is impossible even to count them. This circumstance leaves a certain imprint on sociology. It seems that sociology is still in search of its quintessence.

37. Failures of philosophy of economics

In several axiological branches of science, there is no more authoritative member in methodological terms than economics. Just as in natural science many researchers develop of the philosophy of chemistry and geology in the image and likeness of the philosophy of physics, in social science, when developing the philosophy of axiological branches of science, they often focus on the philosophy of economics. In the world of philosophy of science, it acquired the rights of citizenship earlier than other social sciences, quite successfully assimilating first the achievements of positivism and then critical rationalism.

When characterizing the nature of economics, the principle of maximizing the rate of return on advanced capital is of key importance. There is no doubt that the invention of this principle was the most important achievement of economics over the entire period of its development. The principles of maximization came to economics along with mathematics. The French economist and mathematician Antoine Augustin Cournot used it already back in 1833. Beginning with Adam Smith, economists quite successfully, determining the ratio of economic parameters, mastered the principles and laws of economics. Of course, they had to face difficult questions. This primarily concerned the concept of the value of a product or service. Much to the surprise of economists, it turned out that it is not possible in any way to represent value directly in the form of either a bodily object, or a clot of labor, or a subjective desire. Value is the economic content of such forms of representation of things and processes, which express the most developed economic theories. The value of goods and services and the parameters derived from it cannot be photographed, but it is possible to understand them.

The second difficulty facing economists was the understanding of the scientific method of economics. English economists of the mid-19th century lucky to be contemporary of the eminent philosopher of science John Stuart Mill, the founder of inductive logic. In tying the fate of the method of economics to the method of induction, he was wise enough. Mill understood induction as the processing of facts, finding the causes of phenomena, and the relationships between their parameters.

> If discoveries are ever made by observation and experiment without Deduction, the four methods are methods of discovery: but even if they were not methods of discovery, it would not be the less true that they are the sole methods of Proof; and in that character, even the results of deduction are amenable to them. The great generalizations which begin as Hypotheses, must end by being proved, and are in reality.., proved, by the Four Methods. [Mill, J.S. (2009) A System of Logic Ratiocinative and Inductive. New York: Cosimo, 284.]

Mill adhered to a two-tier conception: induction → deduction, with induction playing a decisive role. Neither abduction nor the cyclical nature of cognition fell into his field of vision. Through induction, he was able to find functional relationships between economic parameters, but not principles. A significant step forward in understanding the method of economics was the views of Alfred Marshall.

> Induction, aided by analysis and deduction, brings together appropriate classes of facts, arranges them, analyses them and infers from them general statements or laws. Then for a while deduction plays the chief role: it brings some of these generalizations into association with one another, works from them tentatively to new and broader generalizations or laws and then calls on induction again to do the main share of the work in collecting, sifting and arranging these facts so as to test and "verify" the new law. [Marshall, A. (1938) Principles of Economics. 8th edn. London: Macmillan, 781.]

In my opinion, Marshall is commendable for two discoveries. First, he understood the equality of induction and deduction. They are equally relevant. Second, Marshall was close to understanding the cyclical nature of cognition, when deduction and induction alternately precede each other. His scheme of conceptual transduction was also two-tier: deduction ↔ induction. This scheme lacked abduction. In its absence, Marshall believed that the broadest generalizations are generated by deduction. In reality, this is due to abduction. As far as I know, none of the economists noticed this mistake.

An important milestone in the development of the philosophy of economic science was the article by Milton Friedman «The Methodology of Positive Economics» (1953). In my opinion, it is a mixture of positivist and critical-rationalistic positions. Positivism manifests itself in Friedman's declaration of economics as a positive science.

> Positive economics is in principle independent of any particular ethical position or normative judgments. As [Neville –V.K] Keynes says, it deals with «what is,» not with «what ought to be.» Its task is to provide a system of generalizations that can be used to make correct predictions about the consequences of any change in circumstances. Its performance is to be judged by the precision, scope, and conformity with experience of the predictions it yields. In short, positive economics is, or can be, an «objective» science, in precisely the same sense as any of the physical sciences. [Friedman, M. (1966) The Methodology of Positive Economics. In Essays in Positive Economics. Chicago: Univ. of Chicago Press, 1966, 4.]

A modern researcher would say that economists over many centuries of the purposeful and at the same time creative work have formed a powerful trend, a kind of attractor. This work is incompatible with various forms of arbitrariness and ill-considered pseudo-innovative proposals. Friedman›s thesis on the positive nature of economics aims to defend this position. Unfortunately, he uses the terminology of physicalism, which is, of course, unacceptable. Contrary to Friedman, no confusion arises when distinguishing between the sciences of nature and society. Nature established the laws of nature, and people developed the laws of economics, gave them the meaning of norms.

One of the features of Friedman›s article is that, without considering the methods of economics, he pays considerable attention to the phenomenon of theory. Here he shifts towards the critical rationalism of Karl Popper. Friedman includes at least four principles of Popper's philosophy in his essays. First, the principle of theoretical relativity: facts are parts of the theory. Second, the principle of falsificationism: facts can never prove a theory; they can only reveal its erroneousness. Third, the principle of the growth of scientific knowledge: any theory is necessarily transient and subject to change with the progress of knowledge. Fourth, the principle of determining the relative strength of a theory: the theory, the conclusions of which are the most accurate, is most effective; the scope of its action is as wide as possible.

Seemingly vigorously trying to understand the nature of economic theory, many economists have shifted their focus from theories to mathematical models. David Houseman claims that

> Most philosophers have argued that science proceeds by the discovery of theories and of laws, but economists are more comfortable talking about models than about laws and theories. ... The really pressing philosophical task for those interested in economics is to come up with an understanding of scientific models, because economic theorizing relies mainly on models. [Hausman, D. (2008) Introduction. In   Hausman, D. (Ed). Philosophy of Economics. An Anthology. Cambridge: Cambridge University Press, 9, 11.]

Most researchers note that models have the resources to manipulate them. Both supporters and opponents of economic mathematical modelling insist on the need to match mathematical models of economic reality.

The emphasis on the conception of a mathematical model of economic reality indicates that many economists are extremely insecure about the conception of economic theory. Mathematical models represent the potential of economic theories; therefore, they in no way cancel the relevance of theories. As for economic reality, it is often seen as something that is originally given, from which models deviate insofar as they use abstractions. Scientists describe reality based on theories including mathematical models. Therefore, there is not and cannot be any contradiction between theories and models, on the one hand, and economic reality, on the other hand. Another thing is that underdeveloped theories and models are not without contradictions. In this case, the illusion is that there is no correspondence between theory and reality. As for abstractions, economists design them for methodological purposes, to simplify theory. They first introduce abstractions and then discard them. The abstractions have a rather indirect relation to the nature of both theories and models.

Thus, the state of affairs in the field of philosophy of economics is not encouraging. First, intratheoretical and intertheoretical transduction is only partially mastered. Second, the frequently encountered denial of the axiological nature of economics is incorrect. Third, it is untenable to oppose mathematical models to economic theories.

38. Language as a representation of theories

Language is one of the forms of presentation of theories. Its two varieties are speech and writing. In the first case, the carrier of the conceptual content of the theory is sound waves, in the second – the material, most often it is paper, on which the signs are applied. Conceptually, language is as rich as theory. The theory is richer than the language in only one respect, namely, the forms of its presentation. After all, theory can be represented not only by language but also by mentality, behaviour, singing, and dancing, as well as by many other forms.

Any science deals with the language expression of its provisions. In this regard, every science is the science of language. This means that not only linguistics is the science of language. Linguistics, however, isolates a certain aspect of the language, in particular, members of speech, for example, a noun and a verb, and members of sentences, for example, a subject and a predicate.

No science can do without linguistics. This means that linguistics is an auxiliary science about all other sciences. Studying the nature of language, linguists have made remarkable strides in the development of grammar, phonology, morphology, syntax, discourse analysis, and many other linguistic sciences. However, in all these cases, no fundamental importance was attached to the decisive circumstance, namely, that linguistics in any of its varieties is a form of representation of a theory. In this case, I use the theory of conceptual transduction, which allows expressing the nature of language much more fully than any other theory.

Criticism of understanding words as designations for things

From time immemorial, many linguists believed that words are designations, signs of what is outside of them, for example, of objects or fragments of mentality. Ludwig Wittgenstein in the late period of his work strove to develop a different conception of language. "For a large class of cases of the employment of the word 'meaning'— though not for all—this word can be explained in this way: the meaning of a word is its use in the language" (Philosophical Investigations, paragraph 43). Now he understands language as an active form of human life. In this regard, it is reasonable to recall also the famous saying of Martin Heidegger "Language is the house of being". With all the differences in the views of Wittgenstein and Heidegger, both held a similar position regarding language, believing that it is the centre of philosophical research. However, their argumentation was mainly metaphysical, but I propose to consider the nature of language based on the theory of conceptual transduction.

All representations of the theory, including language, in their conceptual and methodological content, directly coincide with the theory. A theory does not precede them in any form. Given this circumstance, it immediately becomes obvious that a theory is not a designation for something outside of it. Words are not a designation for things. They have meaning in and of themselves, regardless of things. Words are not signs. The judgments I have made in the light of the centuries-old cultivation of the concept of the secondary nature of language seem striking. It seems obvious that, despite what has been said, there is some form of correspondence between the language and its environment. In this context, the meaning of the term 'correspondence' should be viewed with due care. The conceptions of the theory, one of which is language, are identical in content. They are all equally significant. There is no subordination relationship between them. Thus, there is a natural and organic correspondence between language and other forms of theory presentation, but people cannot violate it. In the sciences, variable correspondence is usually considered, but not organic. Contrary to widespread prejudice, the correspondence between language and other forms of theory presentation is organic and not variable.

Criticism of the triune syntactic-semantic-pragmatic approach

In the light of the theory of conceptual transduction, the shortcomings of the triune syntactic-semantic-pragmatic approach, widely cultivated in linguistics, become obvious. The fallacy of semantics was shown in the previous paragraph. Semantics is built on the notion of a variable correspondence between the language and its environment, which in fact does not exist.

In linguistics, the syntax is the set of concepts that govern the structure of sentences in a given language. In this case, as a rule, special attention is paid to the word order. If we proceed from the theory of conceptual transduction, then it is obvious that the structure of words is determined by its conceptual and methodological content. Linguists do not take into account this circumstance in any way. It does not follow from this that the propositions of synthesists are meaningless. Essentially another, they are significantly depleted in content. This is a consequence of the fact that the source of the content of the sentences is lost, i.e. conceptual transduction.

In linguistics, pragmatics is an addition to semantics. Semantics considers the meaning of linguistic expressions regardless of their situational expression. Pragmatics takes into account the nature of a person's linguistic actions in a specific situation and connection with a specific context, usually determined by the listener. Within the pragmatic approach, there are many concepts, in each of which linguists try to express its true nature. In this regard, for example, Roman Jacobson identifies six functions of the language, the referential, the expressive, the conative, the poetic, the phatic, and the metalingual function. The actual stages in the development of linguistic pragmatics were the concept of the language game by Ludwig Wittgenstein and the theory of speech acts by John Austin and John Searle.

My task is not to consider in detail the various options for linguistic pragmatics. In my opinion, they all have significant drawbacks insofar as language is viewed in itself, and not as a representation of cycles of conceptual transduction. As a result, the content of the language is significantly depleted. Linguistic pragmatists understand language as a form of human practice. In light of conceptual transduction, this conclusion seems both simple and insufficient. Language is as rich as life. In this capacity, it includes practice. Many questions, insoluble from the standpoint of traditional pragmatics, acquire quite a distinct expression in the light of conceptual transduction. Let me demonstrate this with the main difficulty of Ludwig Wittgenstein's concept of language play.

> Here we come up against the great question that lies behind all these considerations. – For someone might object against me: "You take the easy way out! You talk about all sorts of language games, but have nowhere said what the essence of a language-game, and hence of language, is: what is common to all these activities, and what makes them into language or parts of language. So you let yourself off the very part of the investigation that once gave you yourself most headache, the part about the general form of propositions and of language."
>
> And this is true.– Instead of producing something common to all that we call language, I am saying that these phenomena have no one thing in common which makes us use the same word for all,— but that they are related to one another in many different ways. And it is because of this relationship, or these relationships, that we call them all "language". [Wittgenstein, L. (1967) Philosophical investigations. 3rd ed. Trans. G.E.M. Anscombe. Oxford: Blackwell, I 65.]

Wittgenstein believed that all language games have nothing in common. He compared them to board games. A more productive way to clarify the nature of language games is to understand language as a representation of theories. In this case, any language game appears as a clash of theories of the participants in the discourse. Each of them is forced to use the principles and laws of conceptual transduction. This is what all language games have in common.

Linguistics is an auxiliary science. It must be seen as representing theories, primarily scientific ones. Otherwise, its findings are unattractive to scientists. In my opinion, it remains only to regret that until now linguists and philosophers do not use the opportunities that the theory of conceptual transduction opens up for them.

39. Logic as an auxiliary science

As is known, many representatives of analytical philosophy attach paramount importance to the logical analysis of language.

> The business of philosophy, as I conceive it, is essentially that of logical analysis, followed by logical synthesis... The most important part, to my mind, consists in criticizing and clarifying notions, which are apt to be regarded as fundamental and accepted uncritically. As instances I might mention: mind, matter, consciousness, knowledge, experience, causality, will, time. I believe all these notions to be inexact and proximate, essentially infected with vagueness, incapable of forming any part of an exact science. [Russel, B. (1985) [1924] The Philosophy of Logical Atomism. London and New York, 147.]

The obvious successes of analytic philosophy, achieved largely thanks to the pioneering work of Gottlob Frege and Bertrand Russell, have highlighted the exceptional importance of logic, in particular, the logic of first-order predicates. It turned out that logic is not just adjacent to other sciences. It allows you to express their essential features. This means that, as the philosophy of science and linguistics, logic is an auxiliary science. Representatives of any science should be interested in the logical analysis of language insofar as it allows clarifying the conceptual content of the main science. Of course, logical concepts such as logical form and logical sequence are specific. The specificity of logical concepts means that strictly speaking, there is no reason to identify logic with the philosophy of science. The subject of the philosophy of science is conceptual transduction, which does not receive its direct expression in the logical analysis of language. Frege and Russell identified logical and philosophical analysis, but there is no reason for this.

Analytical philosophers distinguish the logical analysis of the scientific and the ordinary language. At the same time, they do not always take into account that both mentioned varieties of language present a certain set of theories. Therefore, it should be about a logical analysis of the theories. In this regard, logic itself acts as a representation of a theory. Representatives of any science themselves can identify logical forms and sequences. Of course, this work is more effective if they use purposefully the achievements of logic. The logic is to summarize the logical achievements of all sciences and make the results obtained at the disposal of each separate science.

The widespread use of the term "logical analysis of language" assigns logical content exclusively to the language. Meanwhile, as I noted earlier, there is an initial, natural correspondence between all representations of the theory. This means that logical analysis is characteristic for all, and not only for language forms of presentation of theories. Instead of "logical analysis of language' it is more correct to speak of "logical analysis of theories".

The logical form, logical consequence, and the logical system as a whole set the specificity of logic. A logical form is a pattern that provides a transition from premises to a conclusion by logical means. Logical consequences are the relationship between statements that hold when one statement logically follows from one or more statements.

A logical system is a conception represented by 1) an alphabet of symbols and expressions composed of them, 2) axioms, 3) formulas (connections between variables and constants), and 4) inference rules that fix the connection between axioms and formulas, as well as between formulas. A formula is a theorem if it is the final link of a logical conclusion. The most important criteria of a logical system are consistency, completeness, and independence of axioms, decidability, and categorically. None of these criteria is absolute. The meaning of each of them depends on the content of the logical system.

Logic, like mathematics and computer science, belongs to the class of auxiliary sciences. This means that this or that logical concept organically expresses certain features of the main theory. Having become interested, for example, in the logical nature of a physical theory, the researcher uses the understanding of its principles, laws, and variables, which is determined by the specific content of this theory. Logic is given the floor after physics, and not before it or independently of it. In interdisciplinary connections, logic is a representation of the main theory. If three-valued logic is used in quantum mechanics, then, in this case, it is a representation of a physical theory. If it is used to clarify the nature of some discourses, then, acquiring new features, it becomes a representation of some logical conception. In addition, one should take into account a variety of logical theories. Jan Łukasiewicz, Stefan Kleene, and Dmitry Bochvar developed various versions of three-valued logic. Ultimately, it turns out that the logical representation of a theory fully expresses its specificity. In logical research, researchers do not abstract from the logical features of the main theory. They abstract from such features of it that go beyond logic.

Special mention should be made of the method of logic. It is widely believed that this method is nothing more than a deduction. By definition, the deduction is a transition from premises to conclusion. This is exactly the case in logic; therefore, the deduction is its method. The above reasoning does not take into account the auxiliary nature of logic. In this capacity, it is relevant for all stages of conceptual transduction, and predictions, and for planning an experiment, and for processing facts, and for updating initial principles, and for updating an obsolete theory, and for building league-theories. It turns out that logical deduction organically matches all methods of conceptual transduction. To avoid misunderstanding, one should not equate logical deduction with deduction as a method of conceptual transduction.

The main point of this essay is that logic is an auxiliary science. In this capacity, logic is called upon, firstly, to generalize the logical achievements of all sciences, and secondly, to make them available to representatives of the main theories. We are talking about a kind of conceptual-logical circle, which implies movement from basic sciences to logic and from logic to basic sciences. If ignoring this movement, then logic ceases to be a scientific doctrine. In modern philosophy of science, the circular nature of logic does not receive the desired vivid expression. Representatives of analytical philosophy, having discovered the logical dimension of all sciences, made a great discovery. However, even they, when it comes to revealing in a clear form the logical contents of specific theories, often are not able to overcome the difficulties that arise.

40. How to understand the nature of mathematics?

Eugene Wigner in 1960 proclaimed in an extensive article the unreasonable effectiveness of mathematics in the natural sciences. Its nature, including such areas as number theory, algebra, geometry, topology, mathematical analysis, discrete mathematics, and category theory, is undoubtedly unusual, nevertheless, it is quite understandable and, therefore, comprehensible. Representatives of all sciences need mathematics. This takes place insofar as mathematics expresses quite definite features of these sciences. Like logic, mathematics is an auxiliary science. All auxiliary sciences function as part of a certain conceptual cycle. In this regard, the following mathematically oriented metamorphoses take place.

First, scientists derive the mathematical content from independent sciences. The result of such extraction is physical, chemical, technical, economic, and several other specific mathematics. Secondly, scientists generalize the achievements of specific mathematics, the result of which is general mathematics. Third, professional mathematicians develop general mathematics in a relatively autonomous mode. Fourthly, representatives of independent sciences use the achievements of general mathematics through interdisciplinary connections. In the process of scientific research, scientists tirelessly implement the cycle: independent science-specific mathematics – general mathematics – advanced general mathematics – specific mathematics – independent science. The implementation of mathematically oriented conceptual cycles involves the cooperation of mathematicians with representatives of independent sciences. If mathematicians do not take an active part in this cooperation, then they, as a rule, consider general mathematics to be an independent science, which it is not. In this case, its nature becomes very mysterious. There is nothing mysterious about the existence of quarks and leptons, which have a whole bunch of unusual properties. In the same way, there is nothing mysterious about the existence of imaginary numbers, points and lines, functions, and functor. Scientists reveal what exists and what does not through cycles of conceptual transduction.

Poor interpretation of the content of mathematical concepts of general mathematics leads to numerous misunderstandings. We can talk about 3 seconds, 3 apples, 3 girls, and, finally, about number 3, which is a concept, for example, of arithmetic as a kind of general mathematics. What is the meaning of the number 3? It expresses some similarity between the triplets of objects listed above. What are these similarities? The theory, which includes the number 3, and in this example, this is arithmetic, express them. In arithmetic terms, all the indicated triplets of objects are identical to each other. Contrary to popular belief, the number 3 as a concept of arithmetic is not the result of some operations of abstracting from the features of the concepts of the main sciences. It is the result of the successful application of the operation of arithmetic modelling to various basic sciences.

At this point, perhaps, it is advisable to clarify the question of the nature of the entities of general mathematics. As you know, the entities of the natural sciences are objects, in particular, particles, fields, and living organisms. The entities of the axiological sciences are people, individuals, and groups of people. What are the entities of mathematics and other auxiliary sciences? What term is appropriate to call them? Scientists have no definite answer to this question. I propose to call the entities of auxiliary sciences uniforms, i.e. forms expressing the identity and unity of the entities of the main sciences.

Of particular note is that mathematics is present at all stages of conceptual transduction. This became clear after the introduction of mathematics to the experimental stage. Imre Lakatos explained the essence of this situation in his book "Proofs and Refutations" (1976). Experiments show the consistency of the concepts of not only natural and axiological sciences but also mathematics. The situation in Einstein's relativistic theory of gravitation is indicative in this respect. Physicists use Riemannian geometry. If Euclidean geometry is used, then, as Hans Reichenbach showed, it is necessary to introduce the concept of universal forces that are not in experiments. The results of the experiments force us to recognize the relevance of not all, but quite definite mathematical theories, which vary from one main science to another.

As for the definitions of mathematical theories, they depend on uniforms and axioms. The degree of completeness of the definition will depend, first, on the number of axioms used in the corresponding definition. My main conclusion is the necessity of interpreting the nature of mathematics in terms of the theory of conceptual transduction. From the same standpoint, it is advisable to interpret the content of the main mathematical directions (programs).

Naturalism in mathematics is Willard Quine's program of developing mathematics through scientific methods. In my opinion, there is no doubt about the relevance of this program.

Formalism in mathematics is a program initiated by David Hilbert to clarify the structure of formal systems. Its decisive point is that in a finite number of moves it is necessary to show the completeness and consistency of the system of axioms of a mathematical theory, as well as its absolute truth. In fact, within the framework of conceptual transduction, mathematicians improve the mathematical theories, but not achieve their absolute truth.

Intuitionism in mathematics is a direction developed by Leitzen Brouwer, in which all objects, except for the initial ones, are the result of mathematical construction. It is believed that the initial objects were chosen intuitively, i.e. without any proof. However, in full accordance with the advances in mathematics, the choice of initial objects varies. This suggests that their definition is not intuitive.

Platonism in mathematics is the recognition of mathematical uniforms as ideal formations that exist independently of people. In fact, along with mathematical theories, uniforms are also improved.

Fictionalism in mathematics is the view, initiated by H. Field, that mathematical objects are abstractions (fictions) and in this capacity contribute to the development of sciences. The auxiliary character of mathematical concepts is correctly emphasized, but they are wrongly identified with abstractions.

Realism in mathematics is the recognition of the reality of mathematical objects. Unfortunately, this reality is not studied in the context of conceptual transduction.

The set-theoretic direction is a program for revealing the unity of mathematics, all its disciplines, based on set theory.

The abundance of mathematical directions shows that pluralism is organically inherent in mathematics. Consistent comprehension of it is a way to overcome mathematical eclecticism. In our opinion, this understanding should be based on the theory of conceptual transduction.

In an essay on mathematics, you should undoubtedly pay due attention to mathematical modelling. Two very energetic scientific parties from opposite positions assess it. Adepts of mathematical modelling believe that it represents the main content of theories. Categorical opponents of mathematical modelling argue that it distorts the perception of many specific features of the studied phenomena. I noted above that the nature of mathematics appears in the context of mathematically oriented conceptual cycles. Then it becomes obvious that successful mathematical modelling reveals the organic mathematical features of the main theories. In this regard, both the exaltation of mathematical modelling and its censure is inappropriate.

41. Computer science as an auxiliary science

All sciences need computer science to a degree no less than mathematics. This circumstance suggests that computer science is an auxiliary science. It deals with conceptual aspects of the world highlighted by using algorithms, programming languages, programs and computers. Those aspects of the world with which other auxiliary sciences deal, in particular, philosophy of science, linguistics, mathematics, and logic, remain outside the attention of computer scientists. Contrary to this circumstance, many researchers believe computer science is a universal science. They absolutize the meaning of computer science. This error usually occurs when the researcher does not consider the correlation of computer science with independent sciences.

Computer science is to a certain extent the result of creative synthesis of the achievements of mathematics, in particular, the teachings of algorithms, and computing. Roughly speaking, computer science is the sum of mathematics and technology. In this regard, a natural question arises about the status that is inherent in computer science. Whether computer science has inherited the formal (auxiliary) status of mathematics or the axiological status of technology. Thinking a lot about this issue, I concluded that computer science as programming is an auxiliary science with a pronounced technological, namely, computer relativity. About technology, programming is an acceptor theory. Technology is its symbol.

Taking into account the auxiliary status of computer science makes it possible to clarify the question of the methods of this science. They are the same as in the independent sciences. One or another kind of computer science goes through all stages of conceptual transduction together with an independent science. Thus, the methods of computer science are deduction, adduction, induction, abduction, updating the principles of the original theory (if possible), and, finally, building a league-theory. Many researchers try to define the methods of general computer science without regard to the methods of the main sciences. Such an operation is doomed to failure. Like general mathematics uniforms, general computer science uniforms do not have space-time characteristics. This means that it is impossible to get facts by operating them. It is widely believed that deduction is the method of general mathematics and programming. However, it is worth pondering over the meaning of deduction. It takes place only in the context of what happens in the cycles of conceptual transduction of independent sciences.

The native place of deduction as a scientific method is in the cycle of intratheoretical transduction. From this form of deduction should be distinguished deduction as a method of general auxiliary sciences, for example, general logic, mathematics, computer science. This time we are talking about a special form of deduction, which I propose to call decomposed deduction. It arises because of the conceptual sublimation of the methods of intratheoretical transduction when their composition is temporarily cancelled.

Since the time of Aristotle and Euclid, scientists developed only the decomposition form of deduction. Its uncritical perception is characteristic of all modern science.

In my opinion, the intratheoretical cycle of the deployment of the content of computer science includes the following stages: paradigms of computations → programs → implementation of programs → evaluation of the results of program implementation → introduction of corrections into the paradigms of computations and the programming paradigm. The content of these steps can be explained by using concepts about the methods of conceptual transduction. Paradigms of computations → programming paradigms → program = deduction. Implementation of the program = adduction. Evaluation of the results of program implementation = induction. Making adjustments in the paradigm of computing and the programming paradigm = abduction.

Many scientists believe that auxiliary sciences do not take into account the peculiarities of the main sciences. This position is in no way consistent with the pluralism characteristic of auxiliary sciences, in particular, for computer science. Computer science deals with the pluralism of paradigms of computations, programming, programming languages, and, finally, programs. Ultimately, only the program that expresses the specifics of the main science turns out to be truly effective. Thus, pluralism is typical for computer science to an extent no less than for other sciences.

When discussing the nature of computer science, the nature of computers raises many questions. Is a computer a physical or technical object? Does he do calculations and think? In my opinion, the computer is not a physical or technical object. We are talking about such a function that it does not have as a physical or technical object. The computer is assigned the function of partially performing the calculation using the processes of magnetization and demagnetization of some fragments of the chips. A person thinks and performs calculations, not a computer. A computer is a computer-scientific object.

In 1950, Alan Turing, raising the question of the similarity of the human brain and the computer, initiated the question of artificial intelligence. Considering the nature of the computer, Turing should have considered its place in the human-computer system. He attributed all the advantages of a human-computer system directly to computers. This was his decisive mistake. As a result, Turing, like all authors who repeat his mistake, awarded computers with properties that they do not possess. Human intelligence is realized primarily in the form of sciences. One of them is computer science.

The issue of computer ethics deserves special attention. There are various conceptions in this regard, two of which are the most popular. According to the first of these, computer technology requires special ethics, which is a modification of traditional ethical systems (James H. Moore and Rafael Capurro). According to the second point of view, informatics has led to a new ethics of a global nature (Luciano Floridi and Krystyna Gorniak). In my opinion, proponents of both concepts do not take into account the status of computer science. The ethical principle is characteristic exclusively for axiological sciences, but not for natural and universally auxiliary ones. Computer science is among the latter. Therefore, as such, the ethical principle is not inherent in it. Computer science is also not a universal science. As such, it lacks the potential for universal ethics. However, computer science is not devoid of ethical horizons. It is necessary for the development of all axiological sciences. In this capacity, it contributes to the development of all varieties of scientific ethics. This means that computer science has ethical relativity. Its content cannot be extracted from traditional ethical systems, in particular, aretological, deontological, and utilitarian ethics. The ethical relativity of computer science is to promote the well-being of all participants in a certain life situation by the principles and laws of those sciences, which it represents.

42. Misunderstood psychology

Contrary to Aristotle, modern authors understand psychology as a theory of not only mentality but also behavioural and language. We are talking about a fairly well-established tradition of understanding the status of psychology, which raises doubts from the standpoint of the principle of theoretical representation. Mentality, language, and behaviour are three of many other representations of a theory, such as an economic or technological conception. Why are only three isolated out of many representations? Why are they registered to psychology? Is it not more expedient to correlate them with the theories of which they represent?

There is a way out of this difficult situation. It consists in considering psychology as a collection of individual theories. In this capacity, it always expresses the specific positions of individuals. The essence of the situation under consideration is that each theory acts as a complex, changeable network of group and individual theories. Individual theories are the prerogative of psychology. All scientists are subjects of psychology insofar as their theories are unique and not reducible to any other conceptions. Einstein's physics differs from Bohr's physics as strictly as Bach's music differs from Beethoven's.

Individual and group theories complement each other. The abundance of individual theories is perhaps the main sign of pluralism of knowledge. At this point, we should remember Michael Polanyi, his bold rejection of the concept of impartial, detached knowledge, to which he opposed personal knowledge.

> ...Personal Knowledge. The two words may seem to contradict each other: for true knowledge is deemed impersonal, universally established, objective. But the seeming contradiction is resolved by modifying the conception of knowing. ...For we cannot possess any fixed framework within which the re-shaping of our hitherto fixed framework could be critically tested. Such is the personal participation of the knower in all acts of understanding. But this does not make our understanding subjective. Comprehension is neither an arbitrary act nor a passive experience, but a responsible act claiming universal validity. Such knowing is indeed objective in the sense of establishing contact with a hidden reality; a contact that is defined as the condition for anticipating an indeterminate range of yet unknown (and perhaps yet inconceivable) true implications. It seems reasonable to describe this fusion of the personal and the objective as Personal Knowledge. Personal knowledge is an intellectual commitment, and as such

inherently hazardous. [Polanyi, M. (1962) Personal Knowledge. Towards a Post-Critical Philosophy. London: Routledge, IV.]

In my opinion, personal knowledge is precisely the world of psychology. The decisive difference between my position and the concept of Polanyi is that I consider personal knowledge as a set of individual theories and certainly in the context of their relationship with group theories. Thus, the definition of psychology as a science of the mentality, language, and behaviour of people is not welcome. Psychology, among other things, really deals with the mentality of people, their language and behaviour, but only insofar as they are characteristic of the individual theories of individuals. Psychologists study thoughts, feelings, emotions, perception, attention, imagination, the motivation of people. They often turn to such phenomena, which are seen at the personal level much more definitely than at the group level, in particular, emotions.

Thus, psychology accompanies all group theories, be they natural, axiological, or auxiliary sciences. This peculiar ubiquity of psychology has repeatedly led many researchers to the idea psychology is the basis of all sciences. The positivists Ernst Mach and Rudolf Carnap, especially the phenomenologist Edwin Husserl, were extremely concerned about this idea. They did not achieve decisive success. Unfortunately, most authors did not compare individual and group theories. The nature of psychology remained largely unclear.

Many researchers have questioned the scientific nature of psychology. The personal character of psychology seemed to be incompatible with the supposedly objective nature of universal laws. On this score, the Simon-Kahneman incident is of considerable interest. As professional psychologists, both researchers became Nobel laureates in economics: Herbert Simon for "decision-making processes in economic organizations" (1978), Daniel Kahneman for "integration of psychological research into economic science, especially concerning human judgment and decision-making under uncertainty" (2002). Both researchers in no way questioned the relevance of scientific methods. At the same time, they noted that economists, using traditional economic theory, did not take into account the personal factors of decision-makers. Most authors believed that Simon and Kahneman introduced psychological research methods into economics. In reality, however, they paid tribute to the individual theories of economic agents, i.e. economic psychology, not psychology as an independent science. Unfortunately, Simon and Kahneman did not consider the cycle of concepts, which unites group and individual economic theories.

A specific psychological component is present in every theory. The generalization of these components leads to general psychology. Psychologists improve general psychology and in this form, return it to group theory. Unfortunately, many psychologists do not consider the origin of general psychology, which they consider independently of special psychology. Their recommendations to representatives of group theories are usually unproductive.

Since, for historical reasons, psychologists have invariably focused on mentality, its nature deserves special consideration. In my opinion, the principle of theoretical representation made it possible to give an adequate explanation of the psychophysiological paradox. Man has the unique ability to present theories in a variety of ways. If this is done through the mechanisms of brain structures, then it is reasonable to speak of mentality. Contrary to Descartes, it is not an independent substance, but a representation of theories through the brain.

Psychologists to clarify the nature of not only thinking, but also feeling have spent much effort. Many researchers believed that the nature of feeling is fundamentally different from the nature of thinking. This assumption was not confirmed. As thinking, feeling acts as a representation of theories. This time it is carried out by other means, namely, emotions, but all the methods of conceptual transduction remain in force. The same methods are characteristic for all representations of theories, including cogitative and emotional ones.

Similar to the state of affairs in other sciences, there is a wide range of directions in psychology. Suffice it to recall in this regard associative, functional, structuralists, gestalt, humanistic, psychoanalytical, behavior, activity, cultural-historical, cognitive psychology. Each of these directions has its own characteristics. Together they give the impression of a chaotic whole spreading in all directions.

However, there is a way to unify them. It consists in interpretation of the content of all psychological directions from the standpoint of the theory of conceptual transduction and the understanding of psychology as a set of individual theories. The very possibility of such an interpretation indicates the unity, diversity, and integrity of modern psychology.

43. Pedagogy at a crossroads

Russian poet Alexander Pushkin began the fifth chapter of his novel "Eugene Onegin" with the words "All of us had a bit of schooling / in something and somehow: / hence in our midst, it is not hard, / thank God, to flaunt one's education". It was said very accurately, but without those very significant obligations expressed by the outstanding American teacher Benjamin Bloom

> Education must be increasingly concerned about the fullest development of all children and youth, and it will be the responsibility of the schools to seek learning conditions which will enable each individual to reach the highest level of learning possible. [Bloom, B.S., Madaus, G.F., and Hastings, J.T. (1981) Evaluation to Improve Learning. New York: McGraw-Hill, Inc., 3.]

Bloom included in the field of education, first, children and youth. Nevertheless, it should include all other segments of the population. All people often without noticing that teach and educate each other. Moreover, they not always properly understand the nature of education. To isolate it, we will have to turn to special science, pedagogy.

Pedagogy deals with a certain type of theories, namely adapted to the conceptual level of learners. Every person and every social group of people use not theories in general, but corresponding to their level of development. Of course, these theories, like all other concepts, are then refined. Pedagogy is to ensure first the selection, and then the possible development of adapted theories. Noted above, apparently, is a sufficient basis for considering pedagogy as an auxiliary science.

Like all other auxiliary sciences, it is to promote the development of independent theories. Pedagogy does not function by itself but as part of a pedagogically oriented cycle of conceptions. In the status of special pedagogy, it functions directly in the basic sciences. The generalization of the achievements of the varieties of special pedagogy leads to the development of a general pedagogy. Educators modernize it and develop recommendations for a specific type of special pedagogy. A significant part of educators makes absolute the importance of general pedagogy. In this case, it is understood not as a component of a pedagogically oriented conceptual circle, but as an independent science. As a result, there is a significant distortion of the scientific nature of pedagogy.

As the author of several dozen pedagogical books, I concluded that the main trouble in general pedagogy is the oblivion of the philosophy of science. The formation of adapted theories without taking into account the achievements of the philosophy of science is invariably accompanied by undesirable incidents. As a rule, they consist of substitution of concepts and methods of philosophy of science with its surrogates. In this regard, the development of a competency-based approach in education is very indicative.

Competence is the readiness and ability of an individual or a team to solve some vital problems. The knowledge, skills, and values acquired in the educational process determine the success of a business. Unforeseen troubles begin when they try to combine the selected competencies and the means necessary for their implementation into a theory. This operation fails due to the special nature of scientific theories. The proponents of the competence-based approach see a way out of this situation in the formation of selected competencies directly in scientifically adapted theories. Meanwhile, theories do not tolerate violence against themselves. It means that the concepts and methods of the theory are precisely true competencies. They do not need doubles or an artificial superstructure over themselves. All genuine competencies ultimately boil down to the concepts and methods of scientific theories. Otherwise, it is impossible to give a consistent interpretation of competencies. Unfortunately, proponents of the competence-based approach, as a rule, do not establish its connection with scientific theories. They are trying to do without the philosophy of pedagogy.

Let us now turn to the question of transforming a more advanced theory into an adaptive conception. The desired goal is usually achieved by the rarefaction of the theory. The skeleton of the theory remains, but the number of concepts decreases. Special attention should be paid to key concepts, for example, the concept of the wave function in quantum mechanics and the concept of value in economics. The historical experience of teachers shows that there is always a successful option for transforming top-scientific theory into an adaptive concept. The main principle of pedagogy is the most effective development of adapted theories of students.

Therefore, according to my analysis, a gross conceptual error is misunderstanding the general pedagogy not as an auxiliary science, but an independent branch of science. This mistake is not accidental; it is the result of oblivion of the philosophy of science and the nature of science. In this regard, the obvious failure of the formation of a holistic approach in education is noteworthy.

The main achievement of humankind is modern science in the unity and diversity of all its more than two dozen branches of science. The exact number of branches of science remains a big mystery. To give the following text the proper degree of certainty, I will assume that there are exactly 20. The education strategy should take into account all 20 basic dimensions of modern science. This is a 20-D strategy (D – dimension). Each person, being a representative of humanity, by definition, must also be a representative of the 20-D education strategy. Formally, this is the case. In all countries of the world, schools and universities study, with rare exceptions, all 20 branches of science. How exactly this done, do is the branches of science under study form a single whole? The corresponding analysis convinced me that there was no trace of the desired unity. Scientific specialization of labour has led to a depressing situation, university professors do not see the scientific whole, and each of them is an expert in only one or two branches of science. In the wake of university professors, the scientific whole also does not get into the field of view of undergraduate and graduate students. As a result, scientific pluralism is accompanied by eclecticism. Can eclecticism be avoided in education? This is possible if not only the general but also the special philosophy of science becomes reliable support of the philosophy of pedagogy, which creates the conceptual framework that is necessary for comprehending science as a whole.

The scientific community is extremely concerned about the lagging pace of education behind the rapid progress of sciences. One gets the impression that pedagogy lags behind other sciences. Without fear of error, we can say with confidence that the mentioned lag is insurmountable if you do not generalize timely the achievements of sciences in the philosophy of science and immediately in the philosophy of pedagogy.

The successful implementation of the 20-D strategy has become a pressing challenge, especially considering the ethical dimension to which the following essay is devoted.

44. The expulsion of ethics from science and the possibility of its scientific rehabilitation

The fate of the ethical project is taking shape in an extremely strange way. The history of this project goes back centuries. Already in antiquity, almost all outstanding philosophers, including Plato and Aristotle, vigorously discussed its content. The ethical project is that advanced knowledge is to ensure the prosperity of all people and not just some of its selected strata. The recent history of human development shows that the ethical project did not materialize. This is unambiguously evidenced, in particular, by the following facts: a fifth of the world's population is malnourished, the same number of people suffer from incurable diseases due to their plight, and this is in conditions when huge resources are spent on the production of weapons of mass destruction.

By the way, is the ethical project relevant? Of course, one can doubt its relevance. In this case, it will be necessary to admit that science is powerless in the face of the main disasters of humankind. This recognition is tantamount to a fiasco of all science, which, of course, is unacceptable. In this case, it is necessary to carry out the project of saving ethics. Whether this salvation is possible is the central question of this essay.

The fate of ethics depends on the history of the development of science and the philosophy of science. The first rather mature phase of philosophy of science was the positivism of Augustin Comte, with its emphasis on facts and the inductive method. Everything else was banished from science, including ethics. There are no ethical facts; therefore, there is no ethics. Ludwig Wittgenstein expressively presented this position was in his famous "Tractatus Logico-Philosophicus" in the framework of the first phase of the development of analytical philosophy, which inherited some of the features of positivism.

> 6.41 [...] In the world everything is as it is and happens as it happens; there is no value in it...
>
> 6.42 Hence there can also be no propositions of ethics. Propositions can express nothing higher.
>
> 6.421 It is clear, that ethics cannot be articulated. Ethics is transcendental.

Wittgenstein's argument is refuted by a simple indication of the ability of people to invent values, including ethical ones. However, this argument seemed dubious to the advocates of impartial positive knowledge. The positivists, as well as their followers from among the analytic philosophers, were clearly at odds with the axiological sciences operating on values.

As for analytic philosophers, many of them have taken a cautious stance in the discussion of the ethical question. For example, George Moor, avoiding superficial judgments, sought to clarify the nature of ethics through a logical analysis of its language, in particular, the language of the ethics of duty of Immanuel Kant and the utilitarianism of Jeremiah Bentham and John Stuart Mill. The results of this analysis did not lead to significant success. The point is that the analyzed theories did not meet the scientific criteria. By itself, a logical analysis of the language of metaphysical systems did not lead to the translation of ethics on a scientific track.

In this context, the case of Bentham is interesting. He stood at the origins of positivism. As a positivist, he did not have to spend efforts on creating utilitarianism, an ethical theory. However, he became its founder. His positivist zeal was manifested in another way, being the founder of not only utilitarianism but also legal positivism, he excluded ethics from the latter.

Positivism-initiated criticism of ethics as a science did not have any significant impact on metaphysically oriented ethical systems. Professional ethicists tirelessly examined various aspects of Aristotle's virtue ethics, Kant's ethics of duty, and utilitarianism. They did not burden themselves with scientific substantiation of the principles of these theories, in particular, Aristotle's principle of "nothing too much", Kant's principle of duty, namely, worthy representation by each person of humanity as a whole, the utilitarian principle of maximum happiness for the largest possible number of people. Other ethical principles were also proposed: the liberation of the proletariat from its exploitation by the bourgeoisie (Karl Marx and Friedrich Engels); the solution of practical problems in a democratic way (John Dewey); the principle of responsibility for the freedom of each individual (Karl Jaspers and Jean-Paul Sartre); the responsibility for the development of technology favourable to people (Hans Jonas); harmonization of ethical values through discourse (Jürgen Habermas), and pluralism of values (Jacques Derrida and Jean-François Lyotard).

The vast majority of professional ethicists were convinced that their ethical recommendations are extremely relevant for all sciences, especially axiological ones, for example, economics, sociology, pedagogy, medicine, and technology. Meanwhile, representatives of the sciences have never rushed to accept the recommendations of ethicists. Many of them sincerely tried to enrich their favourite science with ethical content, but they did not know exactly how to do this.

Let me turn to my theory of returning ethics to the paradise of science. I believe that all attempts to interpret ethics as an independent science are doomed to failure. They were carried out several times, but never led to a union of ethics with sciences. The origins of ethics are not outside the axiological sciences but in them. Deviation from ethical principles often makes itself felt. A demonstrative example in this regard is the state of affairs in the economy.

Economists knowledgeably determine the nature of economic well-being for various segments of the population, including entrepreneurs, employees, and workers. Nevertheless, at the same time, welfare is often maximized in favour of entrepreneurs. An ethical project requires maximizing the welfare of all stakeholders. Neglecting this principle leads to injustice. The path of the desired correction of the discussed situation is obvious: the principles of economic theory should be led by the principle of maximizing the welfare of all stakeholders. This can be achieved only by being guided by the most developed theories. In this, I see the content of the principle of responsibility. Responsibility is not the accountability of the individual to authority, but the desire to ensure the most effective development of the whole, of which individuals or social groups of people are a member.

The example from economics shows that the principles of all axiological sciences should be headed by the principles of maximizing the prosperity of all stakeholders and the principle of responsibility. As for the non-axiological sciences, in particular, natural and formal, they all have ethical relativity. This means that they are essential participants in the interdisciplinary relationship, without which the success of an ethical project is impossible.

Thus, ethics is not an independent science, but an auxiliary conception of all axiological theories. Like all auxiliary sciences, its philosophical companion, i.e. philosophy of ethics, must accompany it. Ethics deserves to be a science. It is time to free it from the snares of metaphysics.

I am far from thinking of underestimating the relevance of metaphysically oriented ethical theories. However, their relevance must be scientifically substantiated. Scientific ethics is the key to understanding metaphysical ethics.

In my opinion, the alienation of ethics from science is the biggest mistake of enlightened humanity in all the centuries of its existence.

45. Three principles of law

Ethics is the science of what should be done, law is the science of what is permissible, exceeding which leads to punishment. Like ethics, the law is an auxiliary science for the entire spectrum of axiological theories. The principles of ethics prevent such realization of the principles of the independent theory, which leads to relapses of mercantile spirit. The law is intended to contribute to the most effective implementation of the principles of both ethics and the core principles of the independent disciplines that determine their specificity, for example, the principle of maximizing profits in the economy or the principle of safe operation of nuclear power plants.

It is widely believed that the law establishes a set of rules, or, in other words, the laws of behaviour of this or that community of people. In this case, the emphasis is on the laws of behaviour and does not take into account that they are always part of a conceptual triad that includes principles, laws, and variables. Auxiliary theories realize their potential as part of independent science and in parallel with its methodological steps. It is always characterized by the same methods of conceptual transduction as for an independent theory. The subjects of law are individuals and social groups of people. Their characteristic legal variables are rights. The relationships between them are legal laws or principles. With this in mind, the definition of law can be as follows: the law is a set of theories in which the management of the concepts of legal principles, laws, and rights is governed through the methods of conceptual transduction. Legal concepts are not imposed on independent theories from the outside. These are concepts of directly independent theories but transformed in a certain way, namely, in such a way as to indicate the limits of their effective functioning. Let me clarify this with an example of economics.

Let us assume that the entrepreneur is guided by the principle of maximizing the rate of return on advanced capital. In this regard, he is forced to endow the subjects of the economy not with any, but with well-defined rights, legal laws, and even legal principles. Qualitative and quantitative attribution of economic concepts turns them into legal concepts. So, if the values of the rate of profit are determined, the achievement of which is accompanied by punishments and rewards, then there is an economic legal principle.

Unfortunately, many researchers think of law not as an auxiliary, but as an independent science. In this case, they need to define the main principle of law, which the Austrian jurist Hans Kelsen called basic norm. Neither he nor anyone else was able to define this basic norm. The fact is that the main legal principle does not exist in its pure form, regardless of the independent theory. He always acts in a certain way transformed the main principle of an independent theory, economic, technological, medical, or some other conception.

The characterization of the nature of law leaves much to be desired. An important milestone in clarifying the nature of the law was Herbert Hart's book "The Concept of Law" (1961). He distinguishes between primary and secondary rules. Primary rules are the total of all legal rules of conduct. Secondary rules mean all legal rules that permit subjects of the right to create, modify, and cancel legal obligations. Moreover, the law is considered an independent science. Philosophically, Hart was guided by the philosophy of the ordinary language of John Austin and Ludwig Wittgenstein.

I understand the nature of law in many ways different than Hart's. In my opinion, the following two circumstances are of decisive importance. First, the understanding of law as an auxiliary science. Second, the interpretation of the nature of legal theories from the standpoint of theories of conceptual transduction. The first circumstance allows us to identify the origins of legal concepts, the second – to express their dynamics. The multiplication of cycles of conceptual transduction leads to the building of the potential not only of independent theories but also of their legal companions.

As you know, a long-term discussion took place about the nature of law between Herbert Hart and his critic Ronald Dworkin. The main object of the dispute was the relationship between law and ethics. Hart believed that the nature of law does not depend on ethics. Dvorkin insisted on the inclusion of ethics in law. In my opinion, the positions of both authors deserve criticism insofar as they isolated ethics and law from those independent sciences about which they are auxiliary concepts. Ethics and law occupy different places in the independent sciences. The subordination of concepts goes from special ethics to the core of the independent theory and further to a special law. Within this line of subordination, ethics and law are separated from each other. However, there is also a definite connection between them. Law is designed to help realize the potential of ethics and the core of the independent theory. Ethics and law in and of themselves have nothing to do with each other. Their relativity is determined by the nature of independent science. This circumstance was not taken into account by both Hart and Dvorkin.

Describing the nature of law, it is desirable, first of all, to highlight its principles. In my opinion, these are the principles of increasing the possibilities of what is permitted, legal responsibility, and punishment. The permissible must be determined, it is not given a priori. It is all the more difficult to determine the possibilities of what is permitted. If the individual does not properly possess the most developed theories, then he will not be able to establish the boundaries of what is permitted. But in this case, he, cannot become a genuine subject of legal responsibility. Legal responsibility is a principle according to which an individual should be guided by the theory of such a level of development that allows him to justify his action. Thus, there is a certain continuity between the principles of increasing the possibilities of what is permitted and legal responsibility. However, they are not deducible from each other, i.e. are principles.

Let us now turn to the principle of punishment. Is it required? The expediency of punishment is confirmed, at least at the present stage of social development, in the process of repeated implementation of cycles of conceptual transduction. Of course, punishment in its meaning is closely related to the principles of increasing the possibilities of what is permissible and responsibility. However, the concept of punishment cannot be derived from these two principles. It adds to the above two principles.

Thus, the legal theory is headed by three principles, the principles of what is permitted, legal responsibility, and punishment.

46. Political science: the war of principles

In the field of axiological sciences, the concept of power is of paramount importance. The development of this concept is political activities. In this regard, they talk about political science. The subjects of political science are political organizations, for example, state and municipal bodies, as well as parties and movements, and individual individuals. It seems that there are all conditions for the recognition of political science as an independent science. This admission is erroneous. The fact is that power becomes valid only if it functions in a certain area, for example, in economics, technology, military affairs, medicine, or education. Вне области своего приложения власть является фикцией. From the above, it follows that about axiological theories, auxiliary sciences are not only ethics and law, but also political science. It is to foster the steady improvement of power to promote the implementation of ethically oriented principles of independent science, such as economics or technology. The main principle of political science is to maximize the effectiveness of power in the context of independent sciences. In this regard, there is a huge range of possibilities, which include, in particular, democracy, authoritarianism, and totalitarianism. Of course, not all of them are acceptable. Only those forms of power relations that are brought into being by the basic principle of political science are approved. Other forms of power relations are denied insofar as they inevitably lead to an unacceptable social hierarchy that degrades the dignity of subordinates and exaggerates the importance of their leaders.

Thus, the principle of the most effective power determines the specificity of political science. However, political scientists very often leave this principle in the shadow of the principles of justice and freedom. The relationship between the three principles is usually not considered. Are the principles of justice and freedom the principles of political science?

In 1971, John Rawls published his monograph "A Theory of Justice", which attracted widespread attention. The main content of the book was the development of the concept of political justice. This concept is developed by representatives of social groups of people who can renounce their interests, as a result, developing a canon of political justice acceptable to all groups of people. At this point, it should be noted that Rawls, wanting to define the status of political theory, ignores its scientific methods. Their status becomes obvious when we consider that political science is an auxiliary science. It uses the same methods as the independent theory, which are nothing but the methods of conceptual transduction. Whatever political justice is, its development is ensured not at the negotiating table, but through the implementation of full-fledged cycles of conceptual transduction. However, how does political justice differ from ethical justice? According to Rawls, political justice is determined not by the distribution of all possible benefits, but exclusively by basic social and economic freedoms.

1. "Each person is to have an equal right to the most extensive total system of equal basic liberties compatible with a similar system of liberty for all".
2. "Social and economic inequalities are to be arranged so that they are both:
   (a) to the greatest benefit of the least advantaged, consistent with the just savings principle, and
   (b) attached to offices and positions open to all under conditions of fair equality of opportunity."

[Rawls, J. (1999) A Theory of Justice: Revised Edition. Cambridge: Belknap Press. p. 266.]

As you can see, Rawls reduces political features to social and economic, and the features themselves to freedoms. After such information, not free from the elements of confusion, it is hardly possible consistently take into account the difference between political science and economics and sociology. Given the status of political science, it is obvious that political justice should be determined in the context of the principle of ensuring the most effective power. The content of this principle is determined by the main principle of independent science, for example, economics, which, in turn, depends on the ethical principle of maximizing the prosperity of all stakeholders. The hierarchy of principles of independent science, for example, economics, appears as follows: the principle of maximizing the economic prosperity of all stakeholders – the principles of economics – the principles of law – the principle of effective power. After all these principles, the question of the status of economic-political justice can be raised, and not about political justice. Moreover, it turns out that there is no need for this principle at all. The fact is that the demand for justice is set 'from above', on behalf of economic ethics, so clearly that it is already contained in the principle of effective power. Like economics, any independent axiological science realizes the principle of justice on behalf of ethics and its principles, and not on behalf of political science.

The freedom advocates Robert Nozick and Friedrich Hayek have most severely criticized Rawls' position. Both consider the fundamental principle of political science not the principle of justice, but the principle of freedom. They believe that abandoning the principle of freedom inevitably diminishes the dignity of people as masters of their own lives. Isaiah Berlin emphasized the existence of two concepts of freedom, negative and positive. Unlike positive freedom, negative freedom presupposes either a complete absence of obstacles to the actions of individuals or their minimization.

In my opinion, the supporters of the principle of freedom insufficient associate its nature with the status of political theory and its principle of maximizing the effectiveness of power. The political scientist finds himself in a position when he should delve into the essence of independent science in detail. In this respect, he can be more or less competent and creative person. In this regard, it is quite natural for him to react negatively to the restriction of his creative scientific activity. Topical is the question not of freedom, but the content-richness of scientific creativity.

Imagine a situation when a researcher seeks to realize the potential of political theory, starting from the principle of freedom. He will not succeed, for this principle does not determine the content of the theory. As you know, the success of theory, and with it the practice, sets only those principles that have the potential to implement its dynamics. There is no way to endow with such potential the principle of freedom. Thus, in my opinion, the principle of freedom is not a principle of political science. The true principle of political science is the principle of maximizing the effectiveness of power. Attempts to oppose it with the principles of justice and freedom do not clarify but distort the nature of political science.

47. Scientific adventures of art history

> Any great work of art ... revives and readapts time and space, and the measure of its success is the extent to which it makes you an inhabitant of that word – the extent to which it invites you in and lets you breathe its strange, special air. [Bernstain, L. (1958) What Makes Opera Grand? Vogue.]

All religions, arts and sciences are branches of the same tree. All these aspirations are directed toward ennobling man›s life, lifting it from the sphere of mere physical existence and leading the individual towards freedom. [Einstein, A. *(2006)* The Einstein Reader. New York: Citadel, *7.]*

Art is the complement of science. Science as I have said is concerned wholly with relations, not with individuals. Art, on the other hand, is not only the disclosure of the individuality of the artist but also a manifestation of individuality as creative of the future, in an unprecedented response to conditions as they were in the past. Some artists in their vision of what might be but is not, have been conscious rebels. But conscious protest and revolt is not the form which the labor of the artist in creation of the future must necessarily take. Discontent with things as they are is normally the expression of the vision of what may be and is not, art in being the manifestation of individuality is this prophetic vision.

[Dewey, J. (1988) The Later Works of John Dewey, Volume 14, 1925 – 1953: 1939 – 1941, Essays, Reviews, and Miscellany. Carbondale, *IL:* Southern Illinois University Press, 113.]

The question of the status of art is overwhelmed beyond measure. In popular classifications of branches of science, art is one of them. However, works in which art is considered a branch of science are extremely rare. Above, I gave judgments about art, which, in my opinion, set the correct perspective for understanding its nature.

What Leonard Bernstein said about art, he could say about science on the same grounds. In terms of ensuring a favourable future for humanity, science and art do not differ from each other. Their nature is clearly of the same type.

Albert Einstein stresses the unity of religion, science, and art. This statement appears to be fruitful. Nevertheless, it raises the question of the nature of this unity. In my opinion, it has been refined throughout the history of human development. In the Middle Ages, science and art were under the auspices of religion. During the Renaissance, the rights of the arts expanded enormously. In our days, science is firmly established in the foreground. This circumstance forces us to consider the nature of art in the context of science. John Dewey expressed this tendency very clearly.

In his opinion, art complements science in two respects. First, it clearly expresses the individuality of the creators of art. Secondly, it expresses the vision of what may be and is not. The first thesis seems to me to be erroneous. The creations of scientists are no less individual than the creations of artists. It is no accident that scientific theories are often named after their creators. The stamp of their individuality marks the theories of Newton, Darwin, Marx, as well as of countless other scientists. As for Dewey's second thesis, it deserves special attention.

The nature of science is the constant improvement of its theories. The same can be said for art. The whole history of its development is also a steady improvement of theories. However, it should be borne in mind that science operates with valid theories. This circumstance is most clearly manifested in the stages of fact-making. Scientists interfere with the real course of events. Unlike scientists, artists do not implement fact-making insofar as they are dealing with fictional. It seems that the decisive difference between art and science has been found. It is so significant that it does not allow enrolling art in the number of branches of science. However, let us not rush to the conclusions.

To understand the relationship between science and art, some theory is needed that would adequately express the nature of both. According to my research, this conception is the theory of conceptual transduction. A far from ordinary circumstance is that, conceptually and methodologically, the theories of science and art are arranged in the same way. In both cases, the same concepts and methods are found. The heroes of novels and operas act in the same vein as real people. They predict events, perform actions, evaluate them in a certain way, and change the principles of their behaviour. However, the aforementioned distinction between science and art remains valid. Art, unlike science, deals with the world of fiction. This circumstance seems to forever-separate science and art. Nevertheless, this impression is deceiving. The fact is that in people's lives they form a single whole. In this regard, people see art as a symbol of their real life. Fiction takes on real meaning. Artistic theories are of the same nature as scientific theories. That is why it is legitimate to include them in the world of science. Of course, artistic theories are special theories, namely, offset conceptions, i.e. theories shifted towards fiction.

Proponents of the contrast between art and science often emphasize that art deals with emotions and science with thoughts. They do not take into account that both thoughts and emotions are representations of the content of theories. Accompanying works of art with a special emotional background does not contradict their scientific nature.

Of particular note are the concepts of principles, beauty, and aesthetics, with the features of which the nature of art is most often associated. Every work of art begins with some idea, which plays the role of a principle. The principle of "Hamlet" by William Shakespeare – maximizing the responsibility of the individual in the face of tragedy. The principle of Dmitry Shostakovich's Seventh Symphony is the victory of genuine culture over its imitation. The principle of Fyodor Dostoevsky's "The Brothers Karamazov" is the authenticity of man in Christ. The composition of a work of art based on a concept becomes a full-fledged deduction.

As for beauty, its nature is connected with the development of theories to an extent no less than the nature of truth. Truth marks the superiority of the more advanced theories over the less advanced ones. It is an integral characteristic of the most advanced theories only. Something similar is the case with beauty. It is an integral characteristic of the most developed art history theories. Moreover, each piece of art is a theory. Any type of art, be it musicology, filmology, literature, or depiction, develops within certain directions. Any attempt to define the nature of art without taking into account the progress of art history theories and trends is doomed to failure. It is in this connection that it is revealed that beauty is an integral characteristic of exclusively the most developed art criticism theories.

According to my research, the identification of the conceptual and methodological nature of art come true in stages. This is, firstly, the development of metatheories, in particular, metamusicology, metafilmology, etc. By definition, the authors of such theories should be the artists themselves. Second, it is necessary to generalize the achievements of all metascientific theories. The result of this generalization is the general philosophy of art. Thirdly, it is necessary to enrich the meta-art theory with the achievements of the general philosophy of art history. The result of this generalization is a special philosophy of art history, in particular, the philosophy of musicology, filmology, theatrology, literature and depiction. Reasonably, the described process of cognition, namely, the trend metaart history – the general philosophy of art history – the special philosophy of art history to call aesthetics. Otherwise, aesthetics degenerates into a set of metaphysical considerations about the nature of art history.

48. History is a complex science

History, unlike all other sciences, has a complex character; it integrates the achievements of all other sciences, considering their variability over time as different components of the past. The past, present, and future are forms not of time, as is often mistakenly believed, but of the variability of phenomena. The past includes those phenomena that have already ended. The present is a collection of ongoing phenomena. The future is phenomena that have not yet begun. Except for history, all sciences deal with the present. The past and the present are explored through the stages and methods of conceptual transduction. The future can be predicted, it is impossible to experiment with it. In the past, all stages of conceptual transduction are contained in the form of some evidence. Historians restore their completeness. Thus, the methods of conceptual transduction are methods of studying not only the present but also the past.

Earlier, when characterizing the specifics of various sciences, I paid special attention to their principles. What principles do historians use? The complex nature of history means that historians have to take into account the hierarchy of the principles of all sciences. In a certain historical period, one or another science dominates, for example, economics or technology. Historians give priority to the principles of dominant science.

The need to consider different types of phenomena in unity forces us to find a single conceptual basis for them. It is physical space and time. The spatial approach allows us to consider phenomena coexisting, in particular, on Earth. The temporal approach allows one to express the sequence of phenomena, of course, taking into account their coexistence. Any consistent historical analysis is characterized by a certain chronotope, the unity of temporal and spatial characteristics of phenomena, no matter how specific their nature is. The Russian literary critic Mikhail Bakhtin developed the concept of a chronotope.

The chronotope as a physical concept is a condition for the existence of non-physical phenomena, but not their characteristic. To express the originality of non-physical phenomena, the historian considers physical spatial and temporal characteristics as carriers of non-physical phenomena, in particular, economic, political, and technological. As a result, he receives a spectrum of relativities of the physical chronotope.

Of course, the nature of non-physical phenomena is not reduced to chronotope, because one or another type of dynamic, in particular, cause-and-effect relations, determines it. Thus, the integration and subordination of different types of past phenomena, including their principles, laws, and cause-and-effect relationships determine the specificity of history.

The philosophy of history is faced with significant difficulties. They are associated with the unwillingness, and the inability of a significant part of historians to integrate diverse social phenomena. Certain stages of conceptual transduction are often unconsciously ignored, which leads to a simplified understanding of the methodological part of the theory. History is often thought of as being as homogeneous as other sciences. In reality, however, history, due to its inherent internal scientific heterogeneity, differs significantly from all other sciences.

As for the methods of history, they are the same for all those sciences that are part of it. These are all the same deduction, adduction, induction, and abduction. The historian's primary task is to reproduce and comprehend the theories that functioned in the past, including their methods. The comprehension of these theories will certainly end with their inclusion in a single set of theories, i.e. to the appropriate league-theories. League-theories combine the past with the present. This circumstance indicates the successive ties of history with other sciences.

Some researchers, considering the relationship between the past, present, and future, understand them as causal factors. Many historians believe that the past is a model for the present. Others attach critical importance to the present. Futurists believe that the true meaning of the present and the past is revealed only when they are viewed from the perspective of the future. The mistake of researchers, which absolutize the past, the present or the future, is to reject the principle of theoretical representation. The true meaning of historical phenomena is clarified in the process of analyzing the relevant theories. In this case, it is revealed that exemplary phenomena can take place both in the past and in the present. Of course, there is no reason to deny their possibility in the future.

Quite often, researchers seek to reveal the ethical meaning of history. Some of them even believe that an ethical turn matured in history. The ethical content of history is expressed with difficulty insofar, as a rule, the ethical content of those sciences that are integrated into a historical whole is not considered. Historical ethics cannot be anything other than the result of combining the ethical content of the sciences integrated into history.

Thus, history is to unite the achievements of all sciences, to present an integral picture of modern science as a whole. Unfortunately, it seems to me that decisive successes have not yet been achieved on this path.

49. On the nature of time and durations

In the previous paragraph, I had to turn to the phenomenon of time. It is rather cryptic in many ways. Physicists have studied the problem of time most thoroughly. Nevertheless, many questions remain unanswered. It is not clear, for example, why time is irreversible and one-dimensional. However, these properties of time may be taken as something given, not requiring explanation. I will focus my attention on two issues, firstly, on the scientific validity of the concept of time, and secondly, on its features outside of physics.

The correct method of studying the problem of time, obviously, is that researchers characterize it by means of scientific theories. Here the first surprise awaits us. In all scientific theories, there is a factor denoted by the symbol t. It means that every process has a certain duration. This is obviously the correct statement. However, is it legitimate to identify the duration of processes with time? In connection with this issue, let us turn to physical theories.

Each physical object has many characteristics, in particular, mass, charge, energy and momentum. The aggregates of these characteristics, for example, masses, do not form some special integral whole. Meanwhile, many scientists believe that the aggregates of the durations of various processes form time as a whole. However, there is no reason to think so. There are no mechanisms, which glue durations into time. Thus, there are durations of processes occurring with objects, but there is no time as such.

What is the nature of durations? They are a measure of the progress of a process from its inception to its end. In this quality, durations do not have any competitors. Of course, the counting of durations of processes depends of scientific theories.

Let us now turn to the question of the specifics of nonphysical durations. It is reasonable to expect that the nature of the durations of various processes is not the same. It would be very strange if the specific processes did not form equally specific nonphysical durations. Based on this assumption, let us assume that, for example, in the physical theory there must be physical, and in the economics economic duration. Turning to physical and economic theories, we find that the same parameter (t) appears in them. How to understand this, should economists be content exclusively with the concept of physical duration, abandoning the concept of economic duration?

Continuing the study of the nature of nonphysical durations, let us pay attention to the following circumstance. Let us assume that the value of the duration of some processes is calculated according to the equations of the physical and economic processes, respectively. In the first case, the duration of processes turns out to be a function of physical parameters, in the second – economic ones. This means that physical durations, being involved in economic processes, acquire economic relativity. There are no economic durations other than physical ones, measured not in seconds, but some other units. Nevertheless, there are physical durations that have economic relativity. Consequently, our conclusion is as follows: in nonphysical processes, physical durations have a type of relativity that corresponds to the specifics of these processes.

50. On the nature of space and extents

Many scientists believe that along with time there is space closely related to it. In the previous paragraph, I concluded that there are process durations, but not time. Considering space by analogy with time, it is reasonable to raise the question of the existence of space along with the extents of physical objects. Turning to physical theories, in particular, to relativistic and quantum mechanics, we find the extent of objects (Δr) everywhere. It is futile to look for traces of physical space in physical theories they are absent there. The nature of physical extents is that they set the boundaries of the coexistence of physical objects, for example various elementary particles and condensed media.

What happens to physical extents if they unite with nonphysical processes? There is ample reason to believe that, along with physical durations, physical extents also have nonphysical relativity. There are no technical, medical and economic spaces as such, but there are physical extents that have medical, technical and economic relativity.

According to my observations, many researchers use the concepts of space and time uncritically. These concepts clearly bear the stamp of metaphysical concepts. In my opinion, it is time to abandon the concepts of space and time. The concepts of not only absolute space and time are inconsistent, but also of any other.

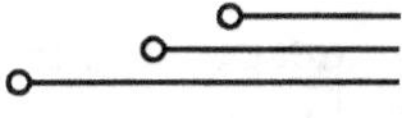

Conclusion

It is time to take stock of my efforts and adventures in the pages of this book. Modern scientists, including philosophers, still misunderstand the actual meaning of the philosophy of science. The philosophy of science is not applied philosophy, but philosophy itself, which is relevant in the scientific sense. The subordination of the philosophy of science to the so-called general philosophy negatively affects the development of all modern science. In this case, it is not understood that the philosophy of science is an auxiliary science, without due attention to which the understanding of the conceptual and methodological structure of sciences leaves much to be desired. Every scientist understands that in the absence of the necessary connection of any science with mathematics, it is conceptually significantly impoverished. To a lesser extent, it is understood that the lack of a proper connection of a particular science with the philosophy of science also leads to its substantial impoverishment.

Insufficient attention to the philosophy of science complicates its formation as a special branch of science. Nevertheless, its formation is taking place. In this regard, it is extremely important to determine the foundations of the philosophy of science. According to my analysis, these are

- the principle of theoretical representation,
- the theory of conceptual transduction,
- a correct understanding of the structure of modern science,
- the main projects of the philosophy of science,
- scientific ethics.

The content of the essays in this book shows that strengthening the positions of all five foundations of the philosophy of science takes place with great difficulty.

Contrary to the principle of theoretical representation, there are widespread attempts to oppose theory to experiment or practice.

Contrary to the theory of conceptual transduction, the internal dynamics characteristic of the theory, as well as the relationships of theories, does not receive its proper expression. Cycles of conceptual transduction are rarely considered.

Contrary to the correct understanding of the structure of modern science, due attention is not paid to auxiliary sciences. They do not include the philosophy of science, linguistics, ethics, law, political science, psychology, pedagogy, and art history.

The main projects of the philosophy of science are its positivist, critical-rationalistic and analytical versions. As a rule, the presentation of their content is not systematic; the provisions of the theory of conceptual transduction are not sufficiently taken into account. Attempts to reveal the content of the philosophy of science from the standpoint of phenomenology, hermeneutics and post-structuralism do not lead to significant results. This means that many philosophers do not so much contribute to the formation of the philosophy of science as harm it.

Ethics is almost entirely at the mercy of metaphysics. Its translation on a scientific track is late. In this regard, the status of science is distorted, its vocation in every possible way to help maximize the prosperity of all people.

I have written this book with a clear desire to bring the philosophy of science to the attention of scientists and other intellectuals. Philosophy of science will establish itself in the status of a branch of science only if the number of its enthusiasts will steadily increase. I urge the readers of this book to be among them.

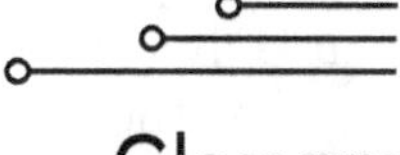

Glossary

Abduction is an adjustment of the principles of a partially outdated theory, taking into account the results of data processing. See duction.

Adduction is a method of obtaining facts during the experiment and observations. See duction.

Deduction is the method of transition from principles to laws and then to variables at the stage of prediction. See duction.

Duktion is the management of concepts or conceptions (theories). Types of duction are deduction, adduction, induction and abduction, and in the case of theories, problematization, discovery, interpretation.

Law is the relationship between variables; at the stage of prediction, researchers deduce laws from the principles by deduction. At the data processing stage, researchers usually identify laws through correlation and regression analysis.

Knowledge is the linguistic and mental representation of the theory.

Induction is a set of methods used in the processing of facts, in particular, dispersion and regression analysis. See duction.

Interpretation is the interpretation of a partially outdated theory at the stage of its renewal through a more developed theory. See duction.

True knowledge is knowledge that is part of league-theories.

Concept is the notion of variables, laws, principles and entities.

Conceptology is the expression of the content of the theory through its concepts.

Conception or theory is a coherent collection of concepts governed by deduction, adduction, induction and abduction.

League-theory is a series of related theories, led by the most developed conception. The content of other theories is interpreted in accordance with the status of the most developed theory.

An example of a league-theory is the electrodynamics of Feynman-Dirac-Einstein-Maxwell.

Metatheory is a scientific theory, accompanied by a special study of its conceptual and methodological content. See protheory.

Method is the same as the duction, way of management of concepts and theories.

Methodology is the expression of the content of the theory through methods.

Observation is a passive experiment, not accompanied by active actions to intervene in the course of events.

Science is a set of league-theories, similar in some respects. Thus, all physical theories are directly related to the principle of least action. All economic theories are involved in the principle of maximization of profit.

Renewal of the partially outdated theory is the stage of intra-theoretic cognition, which leads to the disposal of this theory from its shortcomings.

Processing of facts or data is a stage of cognition, following the actualization and preceding the updating of the principles of the partially outdated theory.

Ontology is an object representation of the theory.

Discovery is the method of inventing such a new theory that allows us to overcome the problems of the theory under study at the stage of its criticism.

A variable is a feature or relation of some entities. Variable are, for example, the temperature of the coolant, the pressure of the steam, and the price of the goods.

Practice is one of the forms of actualization within the framework of axiological theories. It is wrong to contrast the practice of theory, because it is part of its composition.

Prediction is one of the four stages in the management of the concepts of theory; it is the management of the concepts of theory through the method of deduction.

Representation of theory is an expression of its content through some selected concepts. Ontological representation – the content of the theory is expressed by objects and subjects. The linguistic representation of the theory is the expression of its content through linguistic forms, in particular, terms and sentences.

Principle is the initial concept in the prediction process, from which researchers derive laws.

Principle of theoretical representation is a position according to which everything that exists can be judged only based on theories. In this connection, researchers recognize reality is as a representation of theory.

Problematization is a method of finding the shortcomings (problems) of the theory under study at the stage of its criticism. See duction.

Pro-theory is a scientific theory, not accompanied by a special study of its conceptual and methodological content. See meta-theory.

Proto-theory is a theory that precedes an exemplary theory accepted by the scientific community because of its perfection. For example, the physics of Galileo is a proto-theory in relation to Newton's physics. See pro-theory and meta-theory.

Entities are objects and subjects as a concrete unity of variables.

Axiological theories are theories, entities of which are people, either individual subjects or groups of people. All technical theories are axiological.

Natural theories are theories, the entities of which are the objects of nature. Natural theories include physical, chemical, geological and biological conceptions.

Formal theories are theories expressing the equivalence (isomorphism) of several theories. Formal are linguistic, logical, mathematical and philosophical theories.

Theory is the same as the. See conception.

Conceptual transduction is the management of concepts and theories through several ductions.

Inter-theoretic conceptual transduction is the management of theories. In its full form, it is the management of theories through methods of problematization, discovery and interpretation.

Intra-theoretic conceptual transduction is the management of concepts of theory. In its full form, it is the management of concepts through deduction, adduction, induction and abduction.

The fact is information about elementary variables obtained in the course of experiments, in particular, practice.

Philosophy is a set of philosophical theories, each of which expresses the equivalence of the concepts of several theories. In this regard, it is not entirely accurate to say that the philosophy is dealing with general laws and principles. See equivalence of concepts or theories.

Analytical philosophy is philosophy with an emphasis on experiment, induction, logical analysis of language and deduction. W. Quine, H. Putnam, and others developed it.

Hermeneutic philosophy is a philosophy with an emphasis on the concepts of discourse and the consensus achieved in it. V. Dilthey, H.-G. Gadamer and J. Habermas developed it.

Critical-rationalistic philosophy is philosophy with an emphasis on the concept of theory, universal law, deduction and falsification. K. Popper and I. Lakatos developed it.

Marxist philosophy is philosophy with an emphasis on impersonal collective activity. K. Marx developed it.

Philosophy of science is a set of theories expressing the equivalence of meta-theories.

Neo-positivist philosophy is similar to positivist philosophy, but the emphasis is not only on experiment and induction, but also on logical analysis of the language. R. Carnap, M. Schlick and H. Reichenbach developed it.

Positivist philosophy is philosophy with an emphasis on experiment and induction. O. Comte, J.S. Mill, R. Avenarius and E. Mach developed it. See neo-positivist philosophy.

Poststructuralist philosophy is philosophy with an emphasis on concepts of discourse and pluralism of points of view. M. Foucault, J. Derrida and J.-F. Lyotar developed it.

Pragmatic philosophy is a philosophy with an emphasis on rational achievement of the best results in practical activity. Ch.S. Pierce, W. James and J. Dewey developed it.

Phenomenological philosophy is philosophy with an emphasis on the synthesis of perceptions in the process of cognition. E. Husserl developed it.

Existential philosophy is a philosophy with an emphasis on the concept of the authenticity of man, which people express in human freedom. K. Jaspers and J.-P. Sartre developed it.

Equivalence of concepts or theories takes place if there is some form of display between them. For example, linear dependencies, on the one hand, of physical and, on the other hand, of biological variables, can be displayed on each other. In this connection, researchers use the concept of linear dependence in mathematics.

Experiment is active intervention in the course of events.

Ethics is a set of theories, the main principle of which is the maximum prosperity for all people.

Philosophical ethics – ethical theories within the framework of philosophical theories, as a rule, do not take into account the achievements of the sciences.

Scientific ethics – the top of axiological sciences, coupled with the cultivation of the principle of the greatest prosperity.